Life Science
in the Twentieth Century

HISTORY OF SCIENCE

Editors

GEORGE BASALLA
University of Delaware

WILLIAM COLEMAN
Northwestern University

Biology in the Nineteenth Century:
Problems of Form, Function, and Transformation

WILLIAM COLEMAN

Physical Science in the Middle Ages

EDWARD GRANT

The Construction of Modern Science:
Mechanisms and Mechanics

RICHARD S. WESTFALL

Life Science in The
Twentieth Century

GARLAND E. ALLEN

Life Science in The Twentieth Century

GARLAND E. ALLEN
Washington University
St. Louis

John Wiley & Sons, Inc.
New York · London · Sydney · Toronto

Library of Congress Cataloging in Publication Data:

Allen, Garland E
Life science in the twentieth century.

(History of science)
Bibliography: p.
Includes index.
1. Biology—History.   I.   Title.

QH305.A44   574'.09'04   74-31295
ISBN 0-471-02336-1
ISBN 0-471-02337-X pbk.

Printed in the United States of America

10 9 8 7 6 5 4 3 2 1

To My Father

Series Preface

T HE SCIENCES CLAIM an increasingly large share of the intellectual effort of the Western world. Whether pursued for their own sake, in conjunction with religious or philosophical ambitions, or in hopes of technological innovation and new bases for economic enterprise, the sciences have created distinctive conceptual principles, articulated standards for professional training and practice, and have brought into being characteristic social organization and institutions for research. The history of the sciences—astronomy; physics and associated mathematical methods; chemistry; geology; biology and various aspects of medicine and the study of man—consequently exhibits both great interest and exceptional complexity and presents numerous difficulties for investigation and interpretation.

For over half a century an international group of scholars has been studying the historical development of the sciences. Such studies have often called for an advanced level of scientific competence on the part of the reader. Furthermore, these scholars have tended to write for a small specialist audience within the history of science. Thus it is paradoxical that the ideas of men who are professionally committed to elucidating the conceptual development and social impact of science are not readily available to the modern educated man who is concerned about science and technology and their place in his life and culture.

The authors and editors of the *Wiley History of Science Series* are dedicated to bringing the history of science to a wider audience. The books comprising the series are written by men who are fully familiar with the scholarly literature of their subject. Their task, and it is not an easy one, is to synthesize the discoveries and conclusions of recent scholarship in history of science and present the general reader with an accurate, short narrative and analysis of the scientific activity of major periods in Western history. While each volume is complete in itself, the several volumes taken together will offer a comprehensive general view of the Western scientific tradition. Each volume, furthermore, includes an extensive critical bibliography of materials pertaining to its topics.

George Basalla
William Coleman

vii

Acknowledgments

Many people contributed to the preparation of this book, some with critical comments, others with much-needed practical assistance. Thomas Hall, Judith Swayze, and Robert Kohler read the manuscript at various stages. Their suggestions and criticisms were extremely valuable in the shaping of the final version. Dan Lane and Marvin Natowicz helped to organize the illustrations and primary-source material. Earlier versions were expertly typed by Louise Qualls. I am greatly indebted to Donna Bishop for the preparation of the index. I also thank the library staff of the Marine Biological Laboratory-Woods Hole, Massachusetts. Most of the book was written over several summers at that outstanding institution; without the help of Jane Fessenden and her cooperative staff, the task of sifting the vast amount of information involved would not have been possible. Finally, I am especially grateful to my wife who suffered patiently through seemingly endless plans for revision.

G.E.A.

Introduction

DESPITE ITS TITLE, this textbook is less a history of twentieth-century biology as a whole than it is the history of a few selected areas whose growth and development is characteristic of the vast field that has become general biology. I will focus on a few historical questions about the growth of biology between 1890 and 1965 and answer those questions by examples taken from specific biological disciplines.

It is clear that biologists around 1890, 1920, or 1960 were, in each case, devising fundamentally different kinds of theories, with different notions of what was an acceptable explanation of biological phenomena. What were these differences? How did they come about? What factors, internal and external to biology, influenced the shift in explanatory criteria? This textbook will provide some possible answers to these questions while broadly outlining the growth of biological sciences in the twentieth century. The biological disciplines to be treated are not only those that commanded the most attention from biologists themselves during the periods under consideration, but also those that have had profound effects among nonscientists. These areas are embryonic development (notably cell and tissue differentiation), heredity (notably the Mendelian and chromosome theories), evolution (anti-Darwinism, neo-Darwinism, and the synthetic theory), general physiology (physiological control mechanisms), biochemistry, and molecular biology. This selection admittedly ignores highly important and influential fields such as immunology, protozoology, bacteriology, ecology, palentology, animal behavior, psychology, and physical anthropology. Similarly, important movements, impinging on or growing directly out of biology, such as eugenics, genetic engineering, environmentalism, and behaviorism (and their relation to the social sciences), the problem of atomic radiation, and the recent environmental crisis have all been omitted or treated briefly.

These omissions are not meant to suggest that such fields or movements were unimportant, or even in some cases less important, than the ones chosen. But selection is necessary if history is to be anything more than a dry chronicle of "facts," if it is to seek the causes behind events, and if it is to portray the complex interconnections between people and ideas in any epoch. Selecting the few strands that will represent the character of the whole fabric of a past age is the historian's most important and challenging job. There are, of course, many ways to view the history of any period or set of ideas, that is, what gives history as a discipline its richness and its meaning. Undoubtedly, another historian, in studying this period, might have made a different selection, conditioned by different biases, or by a different reading of the data. Yet there is some reality to history, and some selection of events and ideas portray that reality more faithfully than others. Thus I have tried to present what I think are the fundamental issues biology faced as it emerged from the nineteenth into the twentieth centuries. In facing those issues biology fundamentally reoriented itself, a reorientation with which biologists continue to struggle today.

If this book has any lasting merit, however, it will be less in the questions that it answers than those that it raises. Instead of drawing definitive conclusions, I have suggested areas of influence and interplay between people and ideas, between one science and another, or between science and philosophy. No science grows up in a vacuum, and part of the historian's task is to delineate the influences that impinge on science from its social and historial context, and the effects that science has on the society around it. Yet such influences and relationships are often elusive, and parallelisms between two areas do not prove causal connections. Yet, if at least some of those who read this textbook are enticed to investigate more thoroughly the problems and issues that I have raised, the text-book will have served its purpose. History is not static, but the events and our reading of them are constantly changing. Thus the best history tries to phrase issues so precisely that others can go beyond and investigate them more thoroughly.

In the jargon of history of science, a distinction is often made between "internalist" and "externalist" views. "Internal" history of science focuses more on the intellectual history of science—on questions of how certain ideas were originated and transmitted from one scientist or one period of time to another. Its aim is to elucidate the more technical side of the development of scientific ideas (such as, "How did the chromosome theory of heredity develop?" Or, "How did Warburg's theory of biochemical oxidation grow out of his study of respiration in whole cells?"). "External" history of science is more concerned with the interaction between social, political, philosophical, and economic factors (operating within society at large), and it gives scientific ideas during particular periods of history. It seeks to explore the two-way street in which scientific ideas grow out of cultural conditions and, in turn, influence future aspects of culture. This textbook is devoted largely to internalist history. In a volume of this size, one or the other had to be chosen if there was to be any unifying thread to the narrative. In many ways externalist history of science is probably the most meaningful for the general reader. Yet, in a young field such as history of science in general, or history of biology in particular, the internalist view must, in many ways, proceed the externalist. It is necessary to know some of the details of how a science develops before asking how that development was influenced by or, in turn, influenced the society at large.

This book initially was conceived as a way of suggesting some of the patterns of development within biology itself during the twentieth century. However, in the last few years, issues relating biology to society during the present century have come more and more to the forefront. Regretfully, topics such as the mutagenic effects of atomic radiation, chemical and biological warfare (including herbicidal warfare), genetic engineering, population control, the "Green Revolution," social Darwinism, and eugenics (including current controversies regarding IQ, heredity, and race) had to be excluded. An attempt to include them would have necessitated a recasting and rewriting of major portions of the textbook, which was neither possible nor, in the long run, desirable. That is the subject for another textbook to which the brave and optimistic can aspire.

Contemporary Biology: A Profile

The two decades between 1950 and 1970 witnessed an enormous burst of activity in the biological sciences. In genetics, biochemistry, cytology, and molecular biology (among others) answers have been found to questions that were posed as long ago as the middle of the last century. Yet such an outburst of activity is not unique with contemporary biology. Similar periods of rapid and fruitful expansion occurred in the seventeenth century, the eighteenth century, the middle of the nineteenth century, and again during the first two decades of the twentieth century. In each of these periods certain ideas, certain fields of study, and certain methodologies became particularly prominent. For the historian, these dominant themes are important in characterizing the various periods in the growth of biology as a discipline.

Contemporary biology is characterized by several important features. One is the firm belief that all biological problems can ultimately be studied on the molecular level. This view does not maintain that studies at other levels of organization, such as that of the cell, the organ, the whole organism, or the population are of no value. In fact, there is a growing awareness among some biologists that it is equally as important to study these higher levels of organization as it is to study the lower, molecular levels. The view that reduction of a complex biological phenomenon to its simpler components (cells or molecules) is a sufficient explanation has become less prevalent among biologists in the early 1970's. Nevertheless, the revolution in molecular biology in the 1950's and early 1960's emphasized the importance of understanding the molecular basis of biological phenomena before trying to approach the larger, higher-level interactions.

Contemporary biology is also marked by a highly experimental viewpoint. The twentieth century has witnessed the continous attempt of biologists to bring their fields of endeavor within the rigorous domains of experimental analysis. This means not so much the introduction of more refined methods of measurement and description (e.g., use of the electron microscope, or use of protein structure as one means of characterizing a species), as the method by which hypotheses are formulated in such a way as to yield predictions that can be put to experimental test. Despite its emphasis on manipulation, experimentation is as much a way of thinking as it is the performance of some specific operation. In this regard there are two types of hypotheses that can be formulated: testable (subject to experiment) and nontestable (not subject to experiment). The hypothesis that plant seedlings turn green as they germinate because of an inner need to be green is a nontestable hypothesis. It might be correct, but there is no conceivable way to show either that it is or is not. On the other hand, the hypothesis that plant seedlings need sunlight to turn green is testable; we can subject one group of seedlings to light, and another to darkness as they germinate. If both groups turn equally green, then the hypothesis is proven false and can be rejected; other factors can then be tried in a similar fashion. On the other hand, if the group exposed to sunlight turns green and the one exposed to darkness remains pallid, the hypothesis is supported. The strength of the experimental mode of thought, and its accompanying manipulative processes, is that they introduce rigor into the investigation of biological processes. They make it possible to differentiate between alternative ideas. alternative ideas.

The emphasis on experimentation and biochemical analysis has often been taken to imply that modern biologists have repudiated the methods of observation and description that were so prevalent a part of natural history in the past. While some biologists in recent years, in the full flush of one or another molecular success, have spoken condescendingly about "mere observation" or "old-fashioned descriptive methods," these processes remain a necessary foundation for any biological (or scientific) inquiry. Every inquiry begins with one or another observation, whether of an organism in the field, or the characteristics of an enzyme in a test tube. And, in conducting experiments, today's investigators must be as observant for details or anomalies as the most meticulous microscopist or field naturalist was in past generations. It is true, however, that modern biologists have demanded more than observation and description of organisms or their processes. This demand is for the methods of experimental analysis and the attempt to study biological processes on a number of levels, chief of which has been the level of molecular interaction.

Factors in the Emergence of Twentieth-Century Biology

The chief goal of this book is to show how biology as it was practiced in the nineteenth century—in natural history, descriptive and speculative; in physiology, largely mechanistic—was transformed into its twentieth-century mold: experimental, analytically rigorous, and integrative. Both the methods and subject matter that characterize twentieth-century biology were strongly influenced by developments in the physical sciences (i.e., physics and chemistry) during the last half of the nineteenth and first part of the twentieth centuries. Biologists in the 1880's or 1920's continually looked to physicists and chemists for models of how scientific investigations should be carried out. Biologists were aware of the truth of the accusation, often leveled at them by workers in the physical sciences, that their field was not scientific—that they could not rigorously prove their conclusions and that many of their lines of evidence were incomplete and tenuous. Most important, they recognized how few of the accepted generalizations in biology (from Darwinian selection to the biogenetic law and the concept of blending inheritance), could be tested by experiment, the very bastion on which physics and chemistry, in the nineteenth century, had raised their banners of success. These problems all became painfully obvious in the furor surrounding Darwin's work during the last 40 years of the nineteenth century. Arguments and counterarguments among Darwin, his followers, and their opponents brought forth a host of divergent opinion, none of which had been put to any clear experimental test. Of course, experiments had been employed continually in one field of biology—physiology. But in most other areas, except for bacteriology and biochemistry in the last half of the nineteenth century, the application of experiments had been nonexistent or, at best, spasmodic. Prior to 1890, there was no experimental tradition as such in cytology, embryology, evolution, population, and field biology. Even the study of heredity, through experimental breeding, had been more of an art—especially in the hands of practical breeders—than a rigorous experimental science. It was the twentieth century that saw the fanning out of the experimental method into all areas of biology.

The first area to emerge as highly (and explicitly) experimental in the nineteenth century was physiology. In 1847 a group in Berlin known as the "medical materialists"[Ludwing von Helmholtz (1821-1894), Ernst Brücke (1819-1896)] produced a manifesto calling for the reintroduction into biology of physico-chemical methods, by which they meant largely experimental analysis. They isolated organs from the body and subjected them to perfusion experiments (i.e., passed fluids of known composition through the organ via the arteries and veins); they stimulated muscles and nerves with electric currents, recorded the passage of impulses along nerve tracts, and measured optical properties of isolated lenses, retinas, and other parts of vertebrate eyes. They saw the organism as a mechanism, a complex machine whose workings man could unravel with the tools of physics and chemistry. Their approach was reductionist in that it sought to take the organism apart and study its parts in isolation: to reduce the whole to its component parts. Their approach was physico-chemical in that the methods of measurement and analysis they employed were borrowed directly from the physics and chemistry laboratories. They were experimental in that they sought to test their hypotheses with living systems in which they studied only one variable at a time. This experimental tradition was carried on and given greater emphasis in physiology during the latter part of the century by the work of François Magendie (1783-1855), Claude Bernard (1813-1878), John Scott Haldane (1860-1936), and others. Rejecting the oversimplified mechanism of the Berlin School, these workers sought experimental methods for studying larger problems of animal physiology—problems involving the integration of functions of many organs and systems. In particular, the work of Bernard stressed the important role that all physiological processes play in maintaining the constancy of the body's overall internal environment. He saw the proper approach to physiology less as the isolation of organs from the total body processes and more as the study of changes, or lack of change, in the chemical constitution of body fluids. The work of Bernard and others following him indicated that it was possible to be experimental and rigorous without being naively mechanistic.

Of particular significance in developing the experimental approach to biology in the nineteenth century was the school of plant physiology led by Julius Sachs (1832-1897) at Würzburg in the 1870s and 1880s. Using methods of chemical analysis, he studied the effects of ionic properties on plant cells and the role of chlorophyll in photosynthesis, and he carried out a series of ingenious experiments on transpiration (water loss through leaves). Sachs was imbued with the mechanistic views of the Berlin Medical Materialists; he was also highly knowledgeable in modern physical chemistry of the day, being a close friend of the Swedish chemist Svante Arrhenius (1859-1927), originator of the theory of electrolytic dissociation (of ions). Sachs' influence was maintained by his teaching (he taught Jacques Loeb and Hugo de Vries, among others) and through an influential textbook (*Textbook of Botany*, 1868).

The spread of experimental methods from physiology to previously more descriptive areas came first in embryology in the 1880s with the rise of the "developmental mechanics" of Wilhelm Roux (1850-1924). Roux and his followers got their experimentalism most directly from physiology, but continually referred to physics and chemistry as the models of scientific endeavor that all biologists should emulate. From embryology the experimental approach spread to cytology and heredity, and finally to evolutionary theory. Each instance saw a transformation of the kinds of questions being asked and the kinds of methods used to answer those questions, from descriptive and speculative to experimental and quantitative. By the 1930s most areas of biology, except perhaps paleontology and systematics, were loudly claiming new advances because of the use of experimental analysis and the methods of physics and chemistry. Biologists, perhaps protesting too much, continually reminded their readers (and one must presume themselves) that only through the use of physicochemical reductionism, or rigorous experimentation, could sound advances be made. And some, whose fields such as heredity had indeed made enormous strides, could claim that nirvana had been reached: biology was as much a scientific discipline as physics and chemistry.

The admiration that most biologists in the early twentieth century showed toward the physical sciences was always at a certain distance—both in concepts and in time. Most biologists, in fact, knew relatively little physics or chemistry, and what they did know was often what they themselves had learned in school—usually 20 or 30 years earlier. Thus there was always a lag in the kinds of physics and chemistry being practiced in physics laboratory and those that biologists were applying to revolutionize their own disciplines. Only later, in the 1930s and 1940s, did workers with actual training and credentials in the physical sciences begin to enter the field of biology, especially in the area of molecular genetics, and with profound results. Those biologists who carried out the transformation were familiar with the physical sciences they wished to emulate in only the most cursory, second-hand way. Emulation of the physical sciences did not end for biology with the conversion of its most important areas to the experimental and physico-chemical methods. Physics itself was not a static science but, instead, during the very period in which biologists were borrowing its methods, physics underwent an internal revolution of profound significance. The period between 1890 and 1920 saw the discovery of X-rays and the rise of the quantum and relativity theories—initiating a philosophical debate that undercut the very foundations of the classical view of matter and the nature of reality. Atoms were no longer regarded as mechanical bodies—the hard, impenetrable units of Newton or Dalton—but were centers of electrical force with magnetic properties and subatomic particles. Electrons could not be located in specific points in space, but had to be described in terms of areas of probability. And it became more and more clear in chemistry that knowing all the subatomic structure and properties of an element provided no way to predict the properties of any compound in which that element took part. The questions of what was real or what was the nature of matter could no longer be answered in a simple way; in fact, to some, nothing seemed real, and science more and more came to be regarded only as conceptual framework that man imposed on the universe. Atoms were figments, and the attributes of any part in isolation were not the same as the part displayed when interacting with others in a whole. There was no appeal to mysticism in such arguments, only a profound change in what men recognized they could specify as a "true" reality. Physicists departed from the older style, mechanistic philosophy, which saw the universe and all its parts as machinelike, a series of individual parts bumping into each other in mechanical ways. They turned, instead, to a more complex and interactive view of natural phenomena. The whole of something was not knowable from the sum of its parts, and its nature could not be predicted simply from knowing all the attributes of the parts studied in isolation. This newer view of physics found expression in the writings of Ernst Mach (1836-1916), Max Planck (1858-1947), Erwin Schrödinger (1887-1961), and Alfred North Whitehead (1861-1947).

Just after World War I the new viewpoint in physics began to enter biology, again through the portals of physiology, with the work on nervous integration and self-maintenance initiated by Charles Scott Sherrington (1857-1952) and Walter Bradford Cannon (1871-1945), and in blood chemistry by Laurence J. Henderson (1878-1942). Sherrington and Cannon went beyond the simple mechanistic models of earlier neurophysiologists such as Helmholtz, showing that the properties of the nervous system as a whole were not simply the sum of the properties of individual neurons. Henderson, in turn, showed how the buffering capacity (ability to maintain chemical neutrality despite addition of acid or base) of the blood was the function of a complex of interacting systems, whose properties as a buffer were even more remarkable than could have been predicted from the buffering capacities of each individual component. The new philosophy as it emerged in physiology (it had no single name) was expressed also in genetics as opposition to the classical gene concept and its emphasis on discrete particles, or genes, in the germ cells, and in molecular biology in more recent years in work on feedback and integrative processes in enzyme systems.

Thus experimentalism and a mechanistic outlook became prominent in biology between 1890 and 1915, borrowed largely from the physical sciences (through physiology) as they had been practiced in the 1850s, 1860s, and 1870s. In the 1920s a less mechanistic trend became discernible in biology, rejecting the simplistic tendency to reduce all biological phenomena to molecular interactions. This trend, which saw organism more as wholes, as interacting systems, was borrowed also from physics (through physiology)—this time the new physics of the turn of the century.

Philosophy and Biology

Underlying the rise of experimentalism in biology at the end of the nineteenth century was a fundamental philosophical change in the view of reality. This period saw the spread of a philosophical materialism from physiology into most other areas of biological thought, replacing the idealism that had persisted in areas such as embryology, taxonomy, comparative anatomy, evolution, and animal behavior. In embryology idealism showed itself as the preformation theory. In taxonomy idealism showed itself in the doctrine of types and the concept of immutability of species. In comparative anatomy idealism blossomed in the early nineteenth century as the doctrine of types, the idealistic morphology of Cuvier and Owen. In evolution idealism was visible in neo-Lamarckism, the doctrine of orthogenesis and all theories claiming a directionality and purpose (teleology) in evolutionary development. And, in the study of behavior (animal and human), idealism was rampant in the form of anthropomorphism, a strong reliance on instincts to explain the origin of all "basic" behavior patterns, and the idea of a basic "human nature." All this idealism was to give way, first in physiology at midcentury, and gradually in other areas as the new century dawned. To understand the change in philosophical outlook, it will be helpful to define exactly "idealism" and "materialism." It would be hopeless to attempt any comprehensive definitions of terms so widely used (and misused), yet some working definition is necessary for later discussions in this textbook.

Idealism, in its purest form, is the view that in the operation of the world, mind, *ideas*, the abstract concept, is primary, and matter is secondary. The idea precedes the material structure, which reflects the idea and is, in fact, the fulfillment of it. Idealists would claim, for example, that the concept of cat preceded the material form of cats as we see them (in the Divine Plan, perhaps, or in the mind of the Creator). Each material object is an imperfect copy of its ideal type. The ideal type in turn is knowable only by study of the imperfect copies. Ideas are the motivating, organizing causes for material phenomena as we see them. Idealists give primacy to perception instead of to the phenomena themselves, some idealists going so far as to claim that there is no external reality whatsoever, only the world of ideas, of sense percep-tions. The more strict philosophical use of the term idealism is to be differentiated from its colloquial use, which often applies the term "idealist" to any when we say that a person is optimistic or believes the world can be better than it is at the moment. The latter use is much more restricted in meaning, although it is not necessarily inconsistent with the broader use of the term. An example of idealism in science can be found in the pre-Darwinian Platonic view of species as abstract, immutable groups that were formed from the mind of the Creator.

Materialism is the view that matter exists prior to and independent of any sense perceptions or ideas about its nature and organization. Phenomena in the world are all derived from matter in motion, acting according to knowable laws. Most materialists, of whatever school, agree that the uni-verse and its phenomena have an existence independent of man and his perceptions. Theories and ideas about the universe to varying degrees may reflect this true reality, but the material nature comes first. Ideas about that nature come second, and are derived from the prior existing, material reality. Jacques Loeb and other scientists who have claimed that life is nothing more than chemistry of complex reaction systems with their own existence independent of man are examples of the materialist point of view in biology. Since the seventeenth century natural science has taken a more and more materialistic outlook, from the atomism of Boyle and Newton to that of La Mettrie and Trembly in the eighteenth century, to that of Dalton and the Medical Materialists of the Helmholtz school in the nineteenth century. Like the term idealism, materialism also has two uses—a general philosophical use and a colloquial use. In the latter case we often speak of someone who is overly concerned with physical things, commodities, as a "materialist." Throughout this textbook I will use both materialism and idealism strictly in their philosophical, rather than their colloquial senses.

There have been two distinct kinds of materialism in the past century. One is *mechanistic materialism*, which I will hereafter refer to simply as mechanism, or the mechanistic philosophy; the other I will call *holistic materialism* (in philosophy and history this has been called dialectical materialism), and I will sometimes refer to it as holism. Mechanistic materialism holds that the best understanding of any phenomenon comes from studying the individual parts of it that interact. The parts are studied in isolation, and the whole is reconstructed as a sum (and nothing more) of those parts. Mechanists do not see so much the complex interactions between parts, but they strive to characterize each part in and of itself. Of course, practical reality often dictates that in biology, or any science, parts must be studied one at a time if any meaningful information is to be obtained with the methods available. A biologist who studies a single enzyme system or a single neuron is not necessarily a mechanist in philosophy. But, if the biologist works exclusively with isolated systems, paying only lip service to the relation of those systems to the whole, or believes that the whole is knowable merely as the sum of the individual parts, then he or she is a mechanist. Mechanistic materialism has often been associated with the methodology of reductionism. *Reductionism* is the view that the most thorough understanding of any phenomenon occurs when that phenomenon can be broken down—reduced— to its lowest (accessible) level of organization. A reductionist approaching a machine would seek to reduce its operation to a few basic principles of levers or cogs interacting in precise ways. A reductionist approach to a cell would be to break it down to its atoms and molecules. Although reductionism is not associated exclusively with mechanistic materialism, throughout the recent history of biology most reductionists have, philosophically, been mechanistic materialists as well.

Holistic (or dialectical) materialism holds that the study of isolated parts is not the most accurate way to comprehend reality. Holistic materialists do not believe that the whole is equal to more than the sum of its parts in some mysterious way, that is, out of some vital or unknowable force. They do maintain, however, that what is important is not simply the sum total of the individual parts, *but how they interact*. Holistic materialists maintain that one of the characteristics of parts is the nature of their interaction with other parts in the whole, and that, in fact, one cannot know about the part without knowing about its interactions, because they, too, help define its character. Thus, while holistic mechanists do not disparage studying parts in isolation, they also seek to study those parts in the context of the whole to which they belong. For example, a holistic mechanist might study a single nerve cell to determine its responses, characteristics of conduction and the like, but he would not claim that this provides any necessary insight into how the nerve cell operates within the intact organism. Further study of nerve bundles, the nervous system, body fluid composition, and hormonal balance would be necessary before any picture could emerge of now nerves function in their biological (real-life) setting.

The distinctions between materialism and idealism, or between mechanistic and holistic materialism, apply to ways of looking at all of experience, and not just to those areas called the natural sciences. Mechanistic materialism was applied by Descartes and Helmholtz to the psychical as well as the physical world. Dialectical materialism was applied by Marx with great force to the study of society, human relationships, economics, and history. And idealism has formed one way of looking at the world from the Greeks to the present. While we will be concerned throughout this book with these views as they apply to the philosophy of science (and of biology in particular), it is important to keep in mind that in the real world no one, scientist or otherwise, is ever 100 percent materialistic or idealistic, mechanistic or holistic in his thinking. A person who is commonly referred to as a mechanist may think mechanistically a good percent of the time, but he may also have varying degrees of holistic materialism or even idealism in his approach to certain problems. Some people may be inconsistent, thinking both idealistically and materialistically about the same set of problems. Others may think materialistically about one set of problems (e.g., about the natural world outside of man), and idealistically about another (e.g., human learning or socio-political issues).

Biologists and historians alike have often juxtaposed the term mechanism with *vitalism*. Vitalism is a biological form of idealism that sees a qualitative difference between the forces, the laws, governing the function of living as compared to nonliving systems. Vitalists maintain that organisms are animated by a force, by a set of properties that have no counterpart in the inorganic world. The remarkable properties of self-repair or reproduction that characterize living systems are, according to vitalists, the product of forces that science can not explain or investigate. These forces exist prior to the organic processes and they direct these processes. From a philosophical standpoint, vitalism, as a brand of idealism, is opposed to all materialism, holistic as well as mechanistic. Traditionally, the mechanistic materialists have spoken more vehemently and persistently against vitalism throughout the history of biology. But as more and more biologists (especially physiologists) in the 1920s and 1930s began to adopt a holistic approach, they found themselves equally opposed to vitalism as their fellow mechanists. Vitalism has come in many forms; even in the twentieth century it has persisted in several guises. Hans Driesch's "entelechy," Henri Bergson's "emergence" principle, and Pierre Teilhard du Jardin's concept of "mind" all have in common a vitalistic foundation: they see something different—a force, principle, or process—acting in living systems and absent from nonliving systems. It is, in fact, this something different that has been used by vitalists to characterize the qualitative difference between living and nonliving systems.

Conclusion

What caused biologists in the 1890s and early 1900s to lean so heavily toward the methods and examples of the physical sciences? And which biologists made the transition? The answers to these questions are not unrelated. The biologists leaning most toward emulation of the physical sciences were the younger workers, those born after 1865. Their shift came primarily as a reaction to the concerns and methods of the older generation of biologists—in some cases their teachers—who dominated the biological scene in the post-Darwinian period. In seeking alternatives to the descriptive and speculative views of their elders, the younger workers turned their eyes to the physical sciences, which had long been regarded as the most exact (and therfore scientific) of the natural sciences.

To determine what ideas and methods the younger biologists reacted against, it is necessary to understand something of the climate of opinion in late nineteenth century biology. To a very large extent that climate was conditioned by the enormous revolution in biological thinking that grew out of the Darwinian theory. Although other areas of biological thought, such as general physiology, physiological chemistry, and bacteriology, were making considerable strides during the last quarter of the nineteenth century, it was Darwinian thought—the subject of evolution itself, as well as Darwin's mode of explanation—that pervaded and preoccupied the thinking of the majority of the workers who called themselves biologists.

Contents

Life Science
in the Twentieth Century

CHAPTER I

The Influence of Darwinian Thought on Late Nineteenth-Century Biology

D ARWIN'S WORK HAD a profound effect on biology in two quite different directions. One was the enormous interest it generated in, and focus it gave to, the study of animal and plant phylogeny. Once the basic concept of evolution itself became accepted in scientific circles, it became possible, and enormously popular, to trace out phylogenetic histories of all sorts of species. So overriding did the aims of phylogeny become during the last 40 years of the nineteenth century that virtually every other biological discipline, except perhaps general physiology and biochemistry, took second place to, or was actually pressed into the service of evolutionary theory. A second direction was the methodology that Darwin's work (principally *The Origin of Species*) set forth. Darwin wrote in *The Origin of Species* that he had proceeded in a basically inductive way, gathering together vast quantities of data—information—which his general theory of natural selection then encompassed like an umbrella. The idea of natural selection was to Darwin an inductive generalization that gave meaning to a whole host of otherwise disparate facts.

Actually, historical study has shown that Darwin proceeded far less inductively than he maintained, or than he chose to present in his writings. Nonetheless, the image that the Darwinian approach presented to the scientific public in the latter years of the century was one based primarily on induction. More and more, the methodological legacy of Darwin came to place a high value on the large-scale comprehensive type of theory, which brought all questions, all types of problems, into its purview. This legacy became of utmost importance in conditioning the development of a large number of grandiose theories that arose toward the end of the century.

Darwin's methodological influence extended in another, related direction. He drew support for the theory of natural selection from many areas of biological research. Systematics, professional plant and animal breeding, biogeography, comparative anatomy, ecology, and embryology were all areas whose findings were incorporated into his arguments in *The Origin of Species*. By so doing, Darwin gained support for his general idea of evolution by natural selection; he also demonstrated the forcefulness of any theory that can, by a single concept, relate evidence from so many diverse areas of biology. Thus, while the broad generalizations that Darwin drew served as an umbrella under which many seemingly diverse areas of biology could be gathered, the use of evidence from so many areas gave further support to the generalizations. There were two sides to the methodological coin in Darwin's legacy, and each supported the other.

At the same time, the Darwinian theory encountered several problems, both substantive and methodological. The major component of the former was Darwin's assertion, without any direct proof, that small, heritable variations arose and persisted in a population. The whole mechanism of evolution by natural selection rested on this idea because, if the small individual variations that exist in a population are not heritable, selection for or against them would produce no change in the makeup of the population, and no evolution would occur. Consequently, the nature of variations—their origin and inheritance—became a topic of considerable concern during the post-Darwinian period. The major methodological problem of Darwin's work was that the mechanism of natural selection was not at the time directly subject to experimental test. While Darwin himself did make reference to experiments in artificial selection (by plant and animal breeders), he used these experiments only as a model (an analogy) for natural selection. It did not prove that selection in nature operates at all in the same way.

The Science of Morphology

The extent and depth of Darwin's influence can perhaps best be understood in terms of the importance to which the field of morphology rose in the 1860s, 1870s and 1880s. By definition "the study of form," morphology encompassed a number of disciplines that we would consider independent fields of biology today: comparative anatomy, embryology, paleontology, and cytology. Although morphology as an area of biology has had a long history, with changing goals and methods, in the post-Darwinian period it was understood to encompass three major aims.

One was to determine the basic unity of plan underlying the diversity of living forms. In particular, this meant to dissect out the constant in animal or plant form and distinguish it from the temporary or adaptive. It was the conviction of morphologists that living organisms were built on one or at most a few fundamental plans that had undergone modification during their past development. The chief method of discovering such unity was that of comparison. This involved looking at all stages of the life history of the organism from fertilized egg to adult, comparing the corresponding stages of one form with those of other forms (species or more generally related groups). Comparative anatomy, the most highly developed of the comparative sciences of the nineteenth century, traced unity of plan through study of homologous (similarly structured although not necessarily similarly functioning) parts in adult organisms of quite different groups (see Figure 1.1).

A second aim of morphology was to discover the common ancestor, the archetypal form, that related two (or more) divergent groups of organisms, or served as an ancient progenitor of a single modern line. To discover an archetype was to reconstruct from fossil, anatomical, or embryological evidence the imaginary organism that could have given rise to various divergent groups. For example, through studies of larval and adult forms of primitive chordates, segmented worms, and starfish, morphologists tried to trace in various invertebrate phyla the ancestral vertebrate type. One of the most celebrated attempts to deduce a common ancestor was that put forth by the great Darwinian advocate, Ernst Haeckel (1834-1919). In his *General Morphology* of 1866, Haeckel tried to show the common origin of all multicellular organisms from a single, gastrulalike (an early stage in embryonic development) ancestor (see Figure 1.2). The Gastrea was a nonexistent form, and Haeckel's theory was a futile essay in imagination. But it did represent one extreme of a much more general trend within morphology: the search for common links between divergent groups.

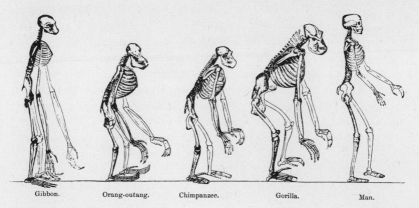

Gibbon.   Orang-outang.   Chimpanzee.   Gorilla.   Man.

Figure 1-1

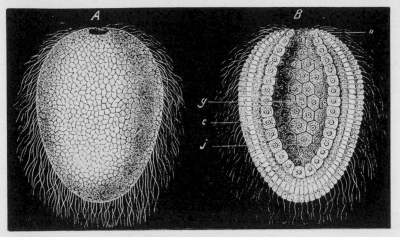

Figure 1-2

A third aim of morphology was the reconstruction of family trees, of phylogenies (the evolutionary course of phyla over time). The evidence for family trees was derived from comparative anatomy, the fossil record, and careful study of embryonic development. For instance, based on the idea of Haeckel's *biogenetic law* (that every organism in its individual embryonic development passes through the major adult stages in the phylogenetic history of its species), morphologists had only to observe how an organism grew from a fertilized egg to an adult in order to see in miniature the sequence of the species' evolutionary past. "Phylogenizing" came to be the dominant concern of many late nineteenth-century morphologists. With overweening zeal these workers constructed family trees with abandon— from those of mollusks and worms to that of man (see, for example, the family tree of man shown in Figure 1.3). The fact that the evidence for many such phylogenetic trees was circumstantial was not a matter of great concern to many morphologists. The basic method of inferring relationships from the study of anatomical and embryological evidence was defended by the great American morphologist W. K. Brooks of John Hopkins University in the following words (1882).

"The evidence for constructing phylogenies is circumstantial, and only leads to general conclusions, and a complete series of fossil forms is the only absolute proof which we could have; but, in the absence of this proof, the conclusions drawn from the study of living animals are rendered extremely probable by the fact that the fossil members of the more modern groups of

PEDIGREE OF MAN

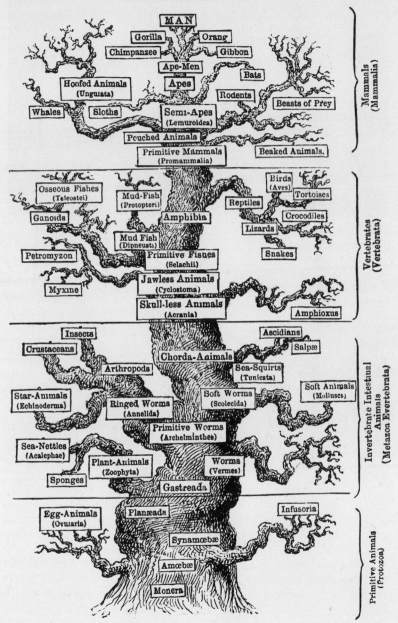

Figure 1-3

animals, such as the mammals and birds, are just such forms as the evidence from other sources leads us to expect, . . . and the attempt to read and interpret such records as we have, and to trace the history of life with as much accuracy as possible, is therefore perfectly legitimate, and may fairly claim the attention of the morphologist."

Concentrating heavily on comparative anatomy, embryology, and paleontology, morphologists paid considerably less attention to the study of physiology. Although most thoughtful workers considered the function of an organ important in understanding its structure, their research drew infrequently on physiological data. For example, organs could be classified functionally as to whether they made contact with the external environment, or whether they did not, but this was ultimately of less concern to the morphologist than classification based on structure. A reason for this lay deep within the morphological tradition. By definition, morphology was concerned first and foremost with form; thus its major methodology was comparative anatomy, with its emphasis on structures, type specimen, or archetypal forms. Functional demands on structures created only modifications, adaptations, or variations of the archetypal form. Thus concentration on function could obscure the underlying pattern of form, since organs known to have no structural relations whatsoever (such as the wing of a bird and wing of a butterfly) could also have the same function. It was clear to most morphologists that functional information was not as basic to understanding the unity of organic types as structural information. Some morphologists, such as the American, C.B. Davenport, were aware of the importance of experimental methods and employed these in their own research. But the majority either ignored physiological methods or, in some, cases, reacted toward them with hostility as being "artificial tampering" with the organism that had no counterpart in nature. Most morphologists saw the organism only as a static being and consequently approached it only with statical methods.

Perhaps the most prominent, if somewhat extreme, proponent of the morphological method was the zealous Darwinian, Haeckel. A naturalist of considerable ability in his early years, Haeckel became a staunch Darwinian after 1859 and spent the better part of his life devising phylogenies for all the major groups of animals and developing his gastraea and recapitulation theories into universal dogma. Through comparative embryology in particular, Haeckel emphasized the importance of studying embryonic development as a means of constructing family trees. His studies began with the fertilized egg and involved detailed microscopic analysis of every subsequent embryonic stage. In the idea of recapitulation, stated as "ontogeny (development of the individual) recapitulates phylogeny (development of the species)," Haeckel claimed that new evolutionary stages of a species were added onto the already present stages of development. Some adult stages were lost in the process, and all were telescoped considerably. But the important feature was the fact that in the embryo Haeckel saw preserved the adult stages of the organism's ancestors. The human embryo, for example, began like the embryo of all vertebrates. It passed through an early stage that was similar for all vertebrates, then developed gill slits and a tail, reminiscent of the fish, then a general mammalian stage in which, among the vertebrates, it was now similar only to other mammals. Eventually development carried the embryo through a primate stage until, as a newborn infant, it had the characteristics only of *Homo sapiens*. Such a series demonstrated to Haeckel that the human species had common ancestry with all chordates, with fish, with all mammals, and with the various primate groups. The temporal sequence showed, in telescoped form, the whole paleontological record of human ancestry. Pushing the argument back farther, the Gastrea was the common ancestor of all multicellular forms: In Haeckel's morphology, there was no need of paleontology and all the ambiguities that the fossil record contained. Armed with his theories and the methods of microscopy, comparative anatomy, and embryology, the morphologist could readily construct the phylogenetic history of any group of animals or plants.

Haeckel was a great publicist and stirred up considerable interest in evolution, embryology, and comparative anatomy. Variations on Haeckel's recapitulation idea appeared in abundance throughout the last decades of the nineteenth century. Although many biologists recognized the recapitulation theory and especially the gastraea theory as oversimplified, they nonetheless found them intriguing. It took many years (well into the twentieth century) to show that embryos themselves adapt through natural selection, and that the developmental stages of any species may well represent the most efficient way of producing adult structures. Of course, closely related species have similar embryonic stages, just as they have similar adult structures and, in that sense, the embryos demonstrate an historical relationship. The fact that at one stage the human embryo has gill slits may indicate only that these structures are a necessary precursor to development of the eustacian tube and inner ear canal, into which they later develop. Such developments may not necessarily indicate anything about direct, common ancestry between man and fish in some remote geological time.

Haeckel's ideas were rivaled for their sweeping generality and popularity by those of his countryman, August Weismann (1834-1914). A cytologist of considerable merit before eye trouble curtailed his laboratory work, Weismann turned his attention in the latter half of his career to the study of problems such as evolution, embryonic differentiation, and heredity. Early on he became a staunch Darwinian. Best known for his doctrine of the separation of germplasm and somatoplasm (the idea that cells of the ovary and testes, which give rise to egg and sperm, cannot be modified by changes in those tissues making up the rest of the body), Weismann was a strong opponent of the idea of the inheritance of acquired characteristics. Changes in the somatoplasm (body cells as distinct from germ cells in the gonads) could not be transmitted from parent to offspring. But Weismann went further than this. In an extremely ambitious conceptual scheme, he attempted to offer a single, unified explanation for evolution, heredity, and embryonic differentiation. To explain heredity, Weismann invented a series of particles that transmitted various characteristics from one generation to the next. To explain how cells became different during embryonic growth, he postulated that these particles were parceled out successively to different cells so that, in the end, each cell had only one kind of hereditary particle—the kind that determined whether that cell was to be muscle, nerve, or skin. To explain how variations arose and were inherited, he postulated natural selection acting on the level of hereditary particles within the germ cells. Competition for nourishment among genetic units allowed some of the stronger ones to gain the upper hand and thus express themselves, while others, the weaker ones, became latent. These were the variations on which natural selection could then act. Weismann used his facts carefully and, ambitious and premature as it may seem to us in retrospect, made an honest attempt at unifying the disparate fields of biology. His work was far more carefully done than that of Haeckel and, in some ways, commanded far more attention from serious biologists who thought Haeckel's ideas flights of fancy. Yet, like Haeckel's, Weismann's theories went far beyond the available facts and as such ultimately came to be regarded by workers in the twentieth century as too morphological and speculative.

The Revolt Against Morphology

It is often a characteristic in the history of science that the major concerns of one generation are often relegated to positions of minor importance by another. In fact, not infrequently the younger workers in a field react negatively toward the ideas and methods of their predecessors and/or teachers. It was in revolting from the domination of much of biology by the aims and methods of morphology that a young generation of workers (that is, born after 1860) created a new analytical and experimental biology in the twentieth century. These younger workers, fed up with the endless pursuit of morphological detail, strove to ask new kinds of questions that called for new methods to answer them. The revolt focused on two aspects of the morphological tradition, which the young workers found stifling.

One was the overriding concern that all aspects of morphology were subservient to problems of evolution. While evolutionary questions were indeed important and interesting, a growing number of biologists felt that problems in other areas, such as heredity or embryology, were being neglected. In fact, specialized studies in those areas, particularly embryology, had been totally subverted to evolutionary ends. Important questions about the process of development and differentiation were neglected as the embryo was studied only for phylogenetic purposes. Similarly, in the pursuit only of evolutionary questions, and with an emphasis on structure, most morphologists were regarded by the younger generation as neglecting functional studies. It was felt that the study of form had totally eclipsed the study of function.

A second aspect of the morphological tradition against which younger biologists revolted was its excessively speculative nature. Haeckel and Weissman were obvious targets, but they were by no means the only ones. It was a characteristic of phylogenizing that the evidence was, as Brooks pointed out, circumstantial. But more than that, morphologists emulated to some extent Darwin's umbrella method of theory formation. To morphologists of Haeckel's or Weissmann's time, no general theory was considered respectable unless it related evolution, embryology, cytology, and heredity. Hence there are innumerable theories between 1870 and 1900 of hereditary units and the role they play in evolution (causes of variation), embryology (differentiation), heredity (continuity between generations), and cytology (chromosomes and other visible cellular structures). To the younger biologists these theories were vapid speculations—they could not be tested and hence could never be proven right or wrong. An infinite number of such theories were possible, but what was the point, the young workers asked, of dreaming up variations and modifications *ad infinitum*? Such pursuits did not bring biologists closer to any real understanding of the world.

To make the problem even more vexing, biologists could not agree even on the question of whether Darwin's theory of natural selection was, in fact, valid. The mechanism of natural selection itself was under considerable fire from many corners of the biological community from 1870 to 1920. In opposition to the neo-Darwinians led by the aged and seemingly indestruc-

tible Alfred Russel Wallace (1823-1913) were a whole host of alternative
schools, for example, neo-Lamarckians, orthogenecists, proponents of
emergent evolution, and mutationists. Each school had its own, largely
speculative theories to account for how one species could be transformed into
another. All of these groups were convinced that the idea of natural selection
could not account for the origin of new species, but none had clear evidence
to support their alternative views. I will say more about one of these
alternatives, mutationism, below.

Given the atmosphere of detailed and endless pursuits of phylogeny on the
one hand, and of unbridled speculation about the causes of evolution on the
other, it is no wonder that younger workers being educated around the turn
of the century turned both to new problems, such as the process of embryonic
differentiation and heredity, and new methods, such as experimentalism, in
seeking their own biological paths. There was nothing of much significance
left to be done in morphology; in fact, even the unanswered questions could
not be seriously tackled with the methods that the morphologists had to
offer. New subjects beckoned with intriguing and fundamental questions.
And new methods—experimentation and rigorous analysis—promised
much, if only they could be used. This was a challenge, implicit for many
and explicit for some, that the younger biologists educated in the 1880s and
1890s met, setting the stage for a wholly new direction for biological
research in the twentieth century.

The extent and significance of the revolt against morphology can be seen
best in a concrete form. Between 1900 and 1910 the "mutation theory" of
the Dutch plant breeder Hugo de Vries (1848-1935) became one of the most
popular and widely accepted ideas within the biological community, particularly among younger workers. The popularity of de Vries' view arose not
only out of the alternatives that he provided to the substance of Darwin's
theory of natural selection, but also out of the experimental methods that he
introduced into the study of a previously descriptive area—evolution. Thus
biologists flocked to de Vries because his theory met many of the objections
to natural selection that had been raised in the years after 1860; they also
found that his methods of crossing and breeding to form new varieties
introduced a totally new kind of investigation into evolutionary studies:
experimentation.

De Vries maintained that new species arose in one generation through the
occurrence of large-scale variations, which he termed "mutations."[1] Darwin
had recognized the possibility of such a process in what he called "sports" or
"monsters," but he had rejected this mechanism as having little or no
significant role in the production of species. De Vries, however, found what

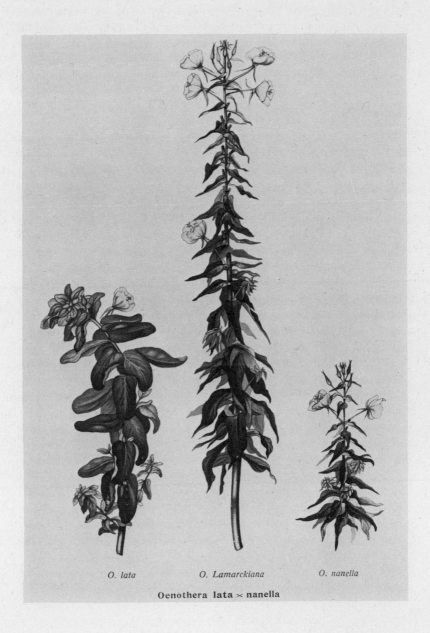

O. lata   O. Lamarckiana   O. nanella

Oenothera lata × nanella

he thought was a clear example of mutations producing species-level differences, and from it he elaborated a large-scale theory that he explicitly hoped
would become a substitute for classical Darwinism. In the 1890s, just
outside the town of Hilversum near Amsterdam, de Vries observed two
apparently distinct strains, or species, of the evening primrose, *Oenothera
lamarckiana,* growing side by side in a field. When self-fertilized[1] each
strain bred true. When DeVries crossbred two strains, generally three
different types emerged that seemed different enough from both the
parents and from each other—in characters such as leaf shape, incision,
variegation, or flower color (see Figure 1.4)—to represent new species.
Through the crossbreeding of *Oenothera,* de Vries considered that he had
found both the appropriate method and organism for demonstrating experimentally how new species could originate in a large, single jump.
Oenothera was just the example that de Vries needed to give substance to his
belief that small, individual variations such as those Darwin had emphasized were not the real basis for the formation of new species.

De Vries identified several kinds of mutations that had somewhat different
patterns of inheritance. Some mutations produced new species at one jump:
the mutation represented enough of a change that an offspring was actually
thought to be of a different species then its parents. Such a mutation might
affect several characters of the organism at a time or only one. Other
mutations were discreet changes, but not of such a magnitude as to create a
new species outright. These variations were generally ones that had occurred
on a larger scale at some time in the past, but had become latent for a period
of time. The accumulation of several such mutations might be necessary for a
new species to be formed.

In *The Mutation Theory,* published in two volumes between 1901 and
1903, de Vries presented mutationism as a general alternative to Darwinian natural selection. The concept that new species originated in one or a
very few steps from preexisting species answered many of the objections that
had been voiced by biologists to the Darwinian theory of natural selection.

First, the mutation theory got around the old argument brought against
Darwin, that newly arising variations would be swamped by the much larger
number of normal, nonvariant forms in the population. Because Darwin had

---

[1] De Vriesian mutation (i.e., =macromutations in today's terminology) were different from the
mutations (=micromutations) commonly referred to in modern biology. Unfortunately, the
same term is used for both, but de Vries' concept postulated much larger single-step changes
than we recognize today. The two ideas had in common only the fact that both types were
considered to be discrete and hereditary.

believed in the idea of blending inheritance, a common conception in the late nineteenth century, he could not explain how new variations, once they arose, would not be diluted in crossbreeding to such an extent that they would become invisible after only a few generations. Since de Vries' theory proposed that new variants were usually already separate species to begin with, and hence could not crossbreed with each other (or if they could, would produce segregating and pre-breeding offspring), the dilution effect of blending could be avoided. De Vries also postulated that mutations occurred in cyclic periods, so that in a period of high activity many individuals might be mutating simultaneously. Thus there might be a number of similar new variants that could crossbreed with each other, offering a further assurance that the new variations would not be lost.

A second problem to which de Vries' theory could address itself was the role of selection. Many critics at the turn of the century attacked Darwinian theory for what they interpreted to be its emphasis on selection as the major factor in evolution. To many, selection was only a negative factor, and they felt Darwin and his followers had avoided the major issue of how species originate by not explaining more clearly the origin of adaptive variations. De Vries could easily dismiss this argument by showing that the new variations arose by mutations and then were acted on by selection. Selection was, indeed, a sieve, eliminating the unfit organism and thus allowing the fit to remain. But it was not, as some Darwinians seemed to say, itself a creative force *causing* the initial appearance of variations.

De Vries' conclusions about the primarily negative effect of selection received considerable support from the experimental studies of the Danish botanist, Wilhelm Johannsen (1857-1927). In 1903 Johannsen showed that selection of continuous variations only separated out the pure lines already existing in a heterogeneous population of organisms. He emphasized that there was a limit to which selection could produce change in a population of organisms. The longer selection was practiced, the less progress could be observed in any of the lines. In selecting for bean weight, for example, Johannsen demonstrated that for the first several generations selection of the heaviest beans in each crop produced an average increase in the weight of the beans in the next generation. However, after a while, increase in average weight per generation slowed down considerably until eventually there was no change from one generation to the next. Johannsen concluded that selection had separated out the hereditary factors for heavy seed from those for light seed and had established a pure line; the pure lines were not, however, new species any more than breeds of dogs or other domesticated animals are new species. Furthermore, Johannsen noted that if selection were relaxed, that is, if a pure line of heavy seeds was allowed to hybridize with a pure line of light seeds, the differences between them would disappear in only one or two generations. Selection thus seemed to produce changes in a population that persisted only as long as the selection process was rigorously maintained. It could not cause the species to transcend that threshold or level of variation between one species and another. Johannsen's results thus fit in well with de Vries' contention that selection does not produce new species, but only separates the fit species from the unfit. The species themselves originated by mutation.

Another argument that Darwin had been unable to answer involved the matter of geologic time. Many critics, including the famous physicist Lord Kelvin, had proposed that the age of the earth was not nearly enough to allow for the evolution of a great variety of animals and plants by slow processes. Perplexed and confused by mathematical and physical arguments, Darwin had admitted quietly that Kelvin presented a real difficulty for his theory. De Vries' Mutation Theory, however, got around this problem by providing a much more rapid kind of evolution. If one new species could arise directly from another in a single generation a great proliferation of species could easily be accounted for within the time span allotted by the physicists for the earth's age.

De Vries' theory could also explain two of the chief anomalies in the paleontological record that had so perplexed Darwin and many of his followers. One, which was always difficult to explain in terms of natural selection, was that fossil forms (of the same or closely related species) seemed to show a regular progression from lower to higher strata, almost as if their evolution was moving in a distinct and straight line, progressive direction. This observation, which seemed to hold for many fossil series, was "explained" by one group of anti-Darwinians by the theory of orthogenesis. Orthogenesis was the idea that organisms varied progressively in a certain direction because of evolutionary momentum. After a few generations of some structure increasing in size, presumably for adaptive reasons, the tendency toward further increase continued, whether such increases were adaptive or not. Orthogenecists pointed to numerous examples of extinction, which were supposed to be the result of a momentum that got out of control, producing exaggerated forms of a character that were nonadaptive. There is no such thing as an evolutionary momentum, of course, and many biologists were unconvinced by the theory of orthogenesis. Still, the paleontological record had to be explained. De Vries' theory offered a much more plausible alternative. He supposed that mutations were more or less random, but that each new mutation was added, cumulatively, onto all those that had preceded it. Progression did indeed occur, but by successful new mutations adding onto the total preexisting characteristcs of the species.

A second anomaly of the paleontological record was the gaps that often existed in related fossil forms between two adjacent strata. Darwin had to explain these gaps—the fact that there were no intergradations between related forms—by imperfections in the fossil record itself. De Vries' theory, however, explained easily why no intermediate forms were found: mutations were discrete and large-scale jumps. The appearance of gaps was only a fossil record of the mutational steps by which species had changed throughout history. The neat explanations that de Vries offered of the geological record was an attractive feature of the mutation theory, since biologists were confused by the apparent lack of clarity with which the Darwinian theory could account for such facts.

A final specific point to which de Vries' theory spoke was the role of isolation in the origin of species. Darwin had been uncertain about whether or not two populations of organisms had to be geographically isolated in order to diverge into separate species. And among his followers the problem of isolation became a burning issue. De Vries pointed out that with the mutation theory the concept of isolation was totally unnecessary. Since a new species arose directly from an old species in one generation, there was no problem of crossbreeding between the new form and the old. Newly mutated species were isolated genetically from their parental form from the start.

All of these specific objections to Darwin's theory, and de Vries' answer to each, were born out of one fundamental misunderstanding among many biologists at the turn of the century: the nature of species. According to our present-day understanding, Darwin himself had erred by arguing that a species was only an arbitrary biological unit created by man for classificatory purposes. A species was thus a taxonomic, but not a biologically functional, unit. In reality, Darwin had been more interested in the evolution of adaptations than in the means by which a taxonomic species evolved. Many followed Darwin either explicity or implicity in this view. Thus the peculiar conclusion had arisen in evolutionary circles by the late 1890s that if species were, in fact, arbitrary groups, it would be futile to try to explain their origin. One of the underlying reasons for the skepticism toward Darwinian theory during this period can be traced to the uneasy feeling among many biologists that Darwin's whole work was based on a logical flaw: it tried to explain the origin of a nonexistent entity.

While many biologists such as the German, Karl Jordan, and the American, W. M. Wheeler, were actively concerned with trying to arrive at a comprehensive view of species as real biological units, the majority of biologists were puzzled, or bored, with the species problem. Many felt that professional taxonomists had reduced the study of species to the building up and breaking down of artificial categories, making it nothing more than a type of post-office activity. Endless debates about whether two insects belonged to the same species or different species convinced younger biologists in the early twentieth century that the whole question of "What is a species?" was meaningless and trivial. This was an unfortunate development in the history of evolution, since it involved biologists in seemingly endless semantic arguments. In many respects the confusion over species contributed significantly to the delay in attempts to formulate a thorough understanding of evolution in terms of populations, which was not to come until the 1920s and 1930s.

In the first decade of the twentieth century the mutation theory was received with enormous acclaim by biologists all over the world. Numerous workers in many fields found that "mutations" were the answer to all the perplexing problems of Darwinian evolution. V. L. Kellogg, an entomologist at Stanford, wrote in 1906:

"On the whole the theory has been warmly welcomed as the most promising way yet presented out of the difficulties into which biologists had fallen in their attemps to explain satisfactorily the phenomena of the origin of species through Darwinian selection."

An even more enthusiastic worker, F. C. Baker, claimed that "no work since the publication of Darwin's *Origin of Species* has produced such a profound sensation in the biological world as *Die Mutationstheorie* by Hugo de Vries." It was in this vein that de Vries had wished—as an alternative to Darwinism—that many biologists view the mutation theory at this time. It seemed to solve all the problems to which numerous other theories, such as orthogenesis or emergence, had spoken only in part.

While the mutation theory had its opponents from the start, strong criticism did not begin to mount until the period between 1912 and 1915. For one thing, although many workers sought them, de Vriesian mutations could not be observed in organisms other than *Oenothera*. For another, the growth of *Drosophila* genetics after 1910 showed that many small mutations occurred in populations of this organism without producing species changes. But probably the most conclusive blow to the de Vriesian theory came with detailed cytological studies (between 1912 and 1915), which showed that the chromosomes of *Oenothera* behaved in very peculiar ways during pollen and egg-cell formation, and that what appeared to be large-scale mutations ment

were actually complex recombinations of already existing characters. In reality, *Oenothera* was a very unusual plant, which led de Vries astray. By 1915 many biologists had abandoned the mutation theory, although de Vries continued championing it until his death in 1935.

The enormous popularity of the mutation theory for over a decade had a further significance other than the substantive arguments it proposed. Many saw in de Vries' work the first application of experimental methods to the nonexperimental and speculative field of evolutionary theory. Trained as a plant physiologist, de Vries had done some important early work on the effects of temperature and osmotic concentration on cell growth and permeability of plant cell membranes. He came to the study of heredity and evolution more or less by chance, and applied to it his concepts of experimentation and laboratory demonstration. De Vries' breeding experiments represented an attempt to control the environmental conditions under which specific hereditary and evolutionary phenomena could be observed. De Vries was distinctly conscious of the importance of bringing experimentation to bear on problems of evolution. As he wrote in the Preface of Volume I of *The Mutation Theory:*

"The origin of species has so far been the object of comparative study only. It is generally believed that this highly important phenomenon does not lend itself to direct observation, and much less, to experimental investigation . . . . The object of the present book is to show that species arise by saltations and that the individual saltations are occurrences that can be observed like any other physiological process. In this way we may hope to realize the possibility of elucidating, by experiment, the laws to which the origin of new species conform."

The most difficult task of evolutionists from Darwin onward had always been to provide definitive proof that species indeed could originate by the action of selection on minute, individual variations. The considerable speculations and grandiose theories that made up the bulk of evolutionary theory reflected this lack of rigorous proof and confounded in a serious way an easy and thorough understanding of Darwinian theory. In response to endless, untestable speculations, workers such as de Vries attempted to demonstrate in a visible way a means by which new species could originate.

That de Vries' introduction of experimentation into evolutionary theory was a determining factor in the enthusiasm with which the work was received is demonstrated explicitly by many of his supporters. The American physiologist Jacques Loeb (1859-1924) praised the mutation theory as introducing into evolutionary theory the same criteria of rigorous experimentalism that he himself had been bringing to the study of animal and human behavior. C. B. Davenport, the American eugenicist, also praised de Vries' experimental approach in 1905, when he wrote (in a review of de Vries' book *Species and Varieties*):

"de Vries' [sic] great work 'Die Mutationstheorie' marks an epoch in biology as truly did Darwin's 'Origin of Species.' The revolution that it is working is less complete, perhaps, because there had remained no such important doctrine as that of continuity to be established. But there was need of a revolution in our method of attacking the problems of evolution. Ever since Darwin's time most biologists have been content to *discuss* and argue on the *modus operandi* of evolution. The data collected by Darwin have been quoted like scriptural texts to prove the truth of the most opposed doctrine. We have seen biologists divided into opposing camps in defense of various isms, but of the collection of new data, and above all, of experimentation, we have had little."

Davenport then went on to say,

"The great service of de Vries' work is that, being founded on experimentation. It challenges to experimentation as the only judge of its merits. It will attain its highest usefulness only if it creates a widespread stimulus to the experimental investigation of evolution."

What the supporters of the mutation theory found valuable in de Vries' work was not merely that it provided a visible demonstration of species formation to which they could eagerly point. It also brought the whole realm of evolutionary theory more closely in conjunction with physiology, and thus ultimately with physics and chemistry. By making the origin of species a functional question—under what conditions could new species originate?—de Vries brought it out of the clutches of mere description, inference, and analogy. The conscious efforts of the times were directed more and more toward squaring all biological theories with physiological concepts and methods.

Conclusion

By the 1890s many biologists working in areas outside of physiology had become exasperated with the aims and practices of morphology. They opposed the dominant role that Darwinian theory had come to play, excluding as it did many other areas (such as embryology) that had an inherent interest of their own. One form of biologists' opposition manifested itself as scientific attacks on weaknesses or ambiguities within Darwin's theory of natural selection. Another form manifested itself as the search for new methods and new areas of concern within the broader field of general biology. New areas included, by the early 1900s, embryology, heredity, and biochemistry. New methods included experimentation and the use of quantitative data.

The new emphasis on experimentation around the turn of the century borrowed much from the view of laboratory science developed in physiology during the nineteenth century. Physiology had many faces during the 1800s, however; the one that was most used as a model by general biologists around 1900 was German physiology. This meant, largely, physiology practiced with a distinctly materialistic and reductivist flavor. It is not coincidental that the early leaders of the revolt from morphology (such as Roux or Driesch) were German, nor that they drew much of their inspiration directly from German physiology of the mid- and late nineteenth century. That view of the proper method of studying organisms which rose to fame with the work of Helmholtz and the Berlin materialsists became the guidelines for a new biology in the early twentieth century.

The revolt from morphology brought a realignment of fields and tradition within the biological community at large. A wedge had long existed beween medically related areas (physiology, physiological chemistry, microbiology,) and natural history (which, dominated by morphology, had included taxonomy, biogeography, evolution, embryology, and comparative anatomy). The new biology saw the movement of embryology, heredity, and eventually evolutionary theory itself from the province of largely descriptive natural history into that of laboratory and experimental analysis. The long-standing separation and distrust between laboratory and field workers was in some ways widened by the growth of strongly experimental sciences from areas that had traditionally been aligned with natural history. In the long run, the naturalist-experimentalist dichotomy, inherited by the twentieth century from a long lineage extending back to the seventeenth century and earlier, has become less pronounced. Although it still persists (sometimes acrimoniously) even today, the introduction of experimental methods into new areas of biology has brought about a fuller integration of biology as a field than ever existed in practice during previous centuries. The synthetic process was long and hard, however, and involved the discarding or modification of old practices, old philosophies, and old prejudices. This struggle forms the central focus of the remaining chapters of this book.

CHAPTER II

Revolt from Morphology I:
The Origins of Experimental Embryology

A S THE DESCRIPTIVE study of embryonic growth and development, the field of embryology reached a high level of achievement by the 1870s and 1880s. The work of classical nineteenth-century embryologists such as T. H. Huxley (1825-1895), F. M. Balfour (1851-1882), Alexander Kowalewsky (1840-1901), Fritz Muller (1821-1897), and W. K. Brooks (1848-1908) generally centered around two types of activities. One was the study of the development of specific structures: the fate of the gill arches in vertebrate embryos, or the origin of the Malpighian tubules in insects, for example. The other was the tracing out of the fate of various germ layers, following precepts laid down in Haeckel's biogenetic law. Both types of activity used observation as their basic methods, and both were concerned primarily with describing events as they occurred in the normally developing embryo. The highly intricate detail and elaborate attention to changes in form from one developmental stage to the next are admirably illustrated in Figures 2.1 and 2.2. As embryos grew, observers noted particularly the changing relationship between the three germ layers; the endoderm (inner layer), mesoderm (middle layer), and ectoderm (other layer). Germ layers are groups of cells in the early embryo from which all the organism's adult tissues later develop (see Figure 2.3). For example, among other things, the endoderm gives rise to tissues lining the digestive tract (innermost tissues in the adult), the mesoderm gives rise to the muscles and bones (middle area tissues in the adult), and the ectoderm gives rise to the nervous system and skin (to some degree, at least, outermost tissues of the adult). The method of formation of the various tissues from their respective germ layers was thought to be (and still is, to some extent) an indication of the species' phylogenetic history, since closely related forms had similar patterns of development. Haeckel was one of the architects of this doctrine; the Germ

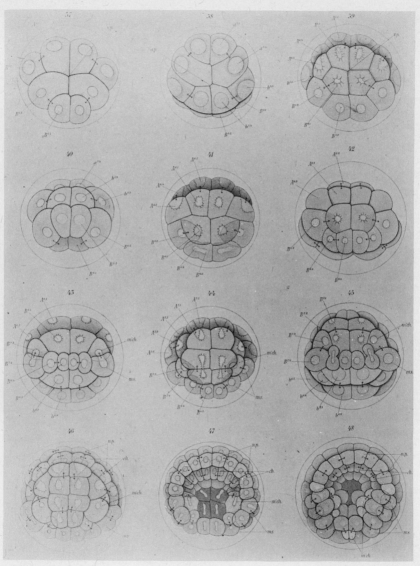

Figure 2-1

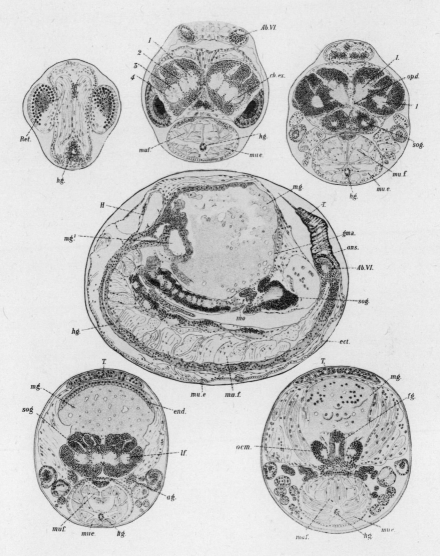

Figure 2-2

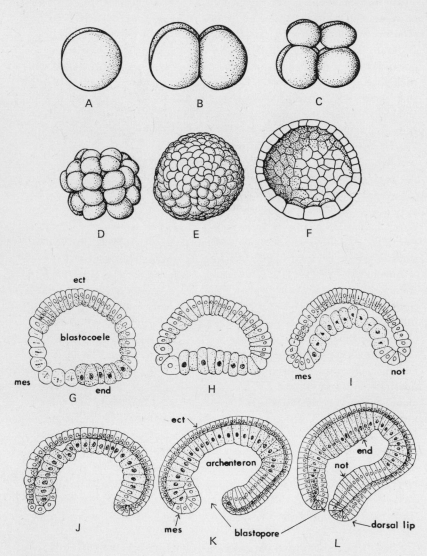

Figure 2-3

Layer Theory became a guiding principle to morphologists as they attempted to chart the topography of the ever-changing embryonic landscape.

The *raison d'etre* of this activity on the part of classical embryologists was, as we saw in Chapter I, grounded in the prevading belief that study of the sequence of embryonic changes threw distinct light on evolutionary history. For example, as a young scientist, Darwin's "bulldog" T. H. Huxley made one of his first contributions to biology by pointing out that jellyfish and their allies consist of only two primary germ layers. Huxley argued in 1849 that these layers corresponded to the ectoderm (outer germ layer) and endoderm (inner layer) of the embryos of higher animals. It was not the existence of the layers in themselves that was of greatest significance to Huxley but, instead, the new method his discovery gave for approaching problems of animal classification and grouping. Even in embryonic development, the search for homologies, which dominated so much of nineteenth-century biology, persisted. Here the homologies were found not only in structures, but also in the very process of development itself. The order of appearance, as well as the origin and fate of particular structures, were all clues on which new phylogenetic trees and new systematic relationships in the animal kindgom and could be based.

But embryology was not to remain for long the playground of morphological speculation. New stirrings were felt beneath the surface persistently from the 1860s onward. The rumblings grew louder, and developed into a full-scale roar when, in 1888, Wilhelm Roux (1850-1924), an embryologist of considerable theoretical grasp and forceful persuasive powers, published the results of his first set of experiments on the embryos of frogs. Roux at one and the same time swept aside the old preoccupation with evolutionary questions and, as an alternative, raised a problem belonging to embryology alone: the mechanism by which embryos grew and differentiated from a formless mass of like cells to an organized group of highly specialized cells. Roux was among the first in the post-Darwinian period to see the embryo as something more than an evolutionary kaleidoscope. He saw the embryo as raising questions of its own, and he devoted the bulk of his life and literary powers to publicizing these questions and what he thought to be the proper methods for answering them. He became one of the chief founders of "experimental morphology" or, as it later came to be known, experimental embryology.

Roux's new departure was born neither overnight nor out of a single set of precedents. Instead, a complex set of factors brought him to the point of breaking with the tradition of descriptive morphology. Some of the factors were peculiar to Roux himself, and his own personal (and largely accidental) history. But others were current in the biological community at large, those historical forces that provide the groundswell from which certain individuals gain their strength and momentum. It is thus important to look behind the development of Roux's experimental approach in order to trace the impetus for this new direction in biology and in order to see clearly the pathway by which experimentalism spread from the physical sciences, through physiology, into embryology.

Roux was not the first to suggest or perform experimental modifications of embryos in order to answer questions about the *causes* of differentiation. Experiments had been used to investigate specific embryological problems in the eighteenth century by Rene Antoine de Reaumur (1683-1757) and Abraham Trembley (1710-1784). These were, however, largely isolated experiments, curiosities of a sort that did not give rise to any lasting tradition of experimental embryology. More profound, however, according to historians, were the nineteenth-century schools that emphasized the physiology of form, an idea championed by Carl Bergmann (1811-1865) and Rudolf Leuckhard (1822-1898), and teratology, the experimental production of monstrosities. Both schools manipulated the embryo in such ways as to produce specific, although not always predictable, structural changes in the adult. Implicit in all of these approaches was the idea that the embryo is a plastic and ever-changing structure. Small changes early in development had manifold consequences in the finished, adult form. The concept of plasticity provided some basis, however tenuous, for thinking about the embryo as something other than the inevitable unfolding of a phylogenetic history. The embryo had its own form and its own responses to conditions; it was not exclusively a slave to its own prehistory.

At the time that Roux was a student, the influence of the experimental embryological work that existed was small. In Roux's own training, were two features that set him off onto a different trend. One was an exposure to the overblown speculations of pre-post-Darwinian era, and the other was the use of experimentation as a tool in other areas of biology, principally physiology and plant cytology. Roux was the son of a fencing master at the University of Jena. In those days Jena was a focal point for important biological research and included on its faculty men such as the physiologist Albert Schwalbe (1844-1916), the botanical cytologist Eduard Strasburger Strasburger (1844-1912), and the brilliant, if controversial, morphologist and Darwinian theorist, Ernst Haeckel (1834-1917). As a student at Jena, Roux studied with these three men and, from their influence, we can trace the seeds of his new approach to embryological questions.

Through Strasburger, Roux was introduced to the importance of seeking general biological explanations on the cellular level. From Schwalbe he

imbibed a strong interest in the relationship between form and function—
that is, the way in which the structure of a cell or organ contributes to its
function and, in turn, the way in which function acts to determine structure.
And, from Haeckel, Roux gained a strong interest in Darwinian theory and
the relation of embryonic development to phylogenetic history. Both
Schwalbe's and Haeckel's work led Roux to the question of the causes of
embryonic differentiation: why do certain cells in the growing embryo begin
at specific stages to become different from other cells? Early in his career
(1869) Haeckel had performed an experiment in which he killed one of the
first two blastomeres (the cells produced by early cleavage of the fertilized
egg) of a growing embryo and looked for the development of abnormalities.
Few, if any, of Haeckel's embryos survived. But the fact that the same
experiment, carried out with greater success, was Roux's most significant
contribution to embryology, suggests the breadth of Haeckel's influence on
his student.

Haeckel's influence on Roux operated in two opposite directions. Roux
admitted that Haeckel directed him toward a mechanistic interpretation of
biological problems, since Haeckel was always talking about physiological
problems and the importance of physico-chemical methods in biology.
Haeckel's philosophy held that causes and their effects were bound intrinsi-
cally to matter in motion. Because the only proper methods of studying
matter in motion were those of the physicist and chemist, Haeckel advocated
the use of experimental techniques in biology wherever possible. The highest
order of business in any science, he felt, was the study of causation; thus, a
proper biologist or mophologist must view the organism as a mechanical and
chemical system. But Roux found that, while he agreed in principle with
this view, Haeckel's own expression of the mechanistic philosophy was too
vague and abstract—too metaphysical—to provide a concrete course of
action. Behind that, however, lay Haeckel's peculiar concept of causation.
To Haeckel, imbued with his biogenetic law, the cause of an embryo's
development was its phylogeny. One stage appeared before another in the
embryo because that stage had appeared earlier in phylogenetic history, and
later stages had simply been added on to the end of development. To
Haeckel, also, there was another sense in which historical sequences displayed causal effects. A later structure was caused by an earlier structure
simply because the latter preceded it in the time sequence of organ formation
in the individual embryo. This is, of course, tantamount to saying today is
the cause of tomorrow because today comes first in time. Neither of these
uses of the term cause was satisfactory for Roux. He sought something more
concrete and direct.

Roux ultimately broke from the tradition of Haeckelian morphology in the early 1880s. Not only was he skeptical of Haeckel's generalizations and abstract ideas, but he was also attracted to the growing body of experimental plant and animal physiology that emerged during the mid- and later nineteenth century. In particular two men influenced Roux's thinking in this direction: Wilhelm Preyer (1842-1897) and Wilhelm His (1831-1904). They turned the general mechanistic bias he had inherited from Haeckel toward concrete examples of the use of experimental and quantitative analysis.

Preyer, a student of du Bois-Reymond and the Helmholtz school in Berlin, was interested in studying the physiology of embryos. Although his concerns were ultimately more physiological than embryological, he analyzed chemical changes occurring during embryogenesis, and thus provided a wholly new way for embryologists to look at their subject. What Preyer showed was that functional as well as anatomical changes occurred during embryonic development, and that these could, to varying degrees, be correlated with each other. Roux studied under Preyer at Jena, attended his lectures in the University's Medical Faculty, and was strongly influenced by the strides that Preyer had made in developing techniques of measuring physiological changes in embryogenesis.

Wilhelm His was a physiologist and anatomist by training who became interested in the physical and chemical forces that caused movements in specific embryonic cells during development. (It is known, for example, that whole groups of cells migrate from the outside of the embryo to the inside during formation of the gastrula stage.) To understand these forces better, His devised a number of mechanical models to show how various stresses and strains could cause folding, bending, and specific cell movements associated with embryonic growth. His' work, unlike that of Preyer, was concerned with problems specifically in the domain of embryology. Although Roux never studied formally with His, as a student he became acquainted with the older man's work, and he was greatly influenced by him to seek causes for embryological events in mechanical and physical terms.

The influences that led young Roux to become the chief exponent of mechanistic and experimental embryology were thus part of a developing methodological tradition within nineteenth-century biology. Originating in physiology, these methods were applied by Roux to the problems that he considered foremost in embryology at the time: the causes of differentiation. Through his strong emphasis on experimental results, and through his strong advocacy of experimentalism as an important—indeed, crucial—method of research, he revolutionaized the field of embryology in particular and of biology in general.

The Mosaic Theory of Development

By 1885 Roux had formulated in his mind a concept of heredity and development that provided a mechanism for embryonic differentiation: the mosaic theory. Most of his experimental work was designed to test, ultimately, one or another aspect of this idea. The mosaic theory held that hereditary particles in the cell were divided in a qualitatively uneven fashion during the cell divisions (called cleavage), which form the multicelled embryo from the single-celled egg. At each division, the two daughter cells would end up with different hereditary potentialities. As cleavage continued, the potentialities of individual cells became more and more restricted; ultimately, a cell would express only one major hereditary trait, being recognized as belonging to one particular tissue type. The beauty of this formulation was that it led to predictions that could be tested. If the hypothesis were true, then destruction of one blastomere at the two- or four-cell stage, for example, should produce a deformed embryo. If the hypothesis were not true, then destroying a blastomere should produce little or no effect. Roux set out to perform the appropriate experiment.

With a hot, sterilized needle Roux punctured one of the blastomeres in the two-cell stage of a frog embryo. The punctured cell was killed, but the other blastomere was allowed to continue development. Roux's results showed that all of the embryos developed abnormally. That is, they lacked some particular set of embryonic parts, and usually failed to develop beyond the late gastrula stage (see Figure 2.4). In essence, Roux got half embryos that, when studied microscopically, showed the cells on one side well-developed and even partially differentiated, while those on the other side were highly disorganized and undifferentiated. Roux interpreted these results as support for the mosaic theory. When its sister cell was killed, the remaining blastomere was unable to develop into a whole embryo because, according to the mosaic theory, it contained hereditary particles for only half of the adult organism.

In his enthusiasm for the mosaic theory, Roux could easily accommodate a set of contradictory experiments published in 1891 by another German, Hans Driesch (1867-1941). Working at the Zoological Station in Naples in the 1880s and 1890s, Driesch had tested Roux's theory with eggs of another organism, the sea urchin. Instead of killing one of the first two blastomeres with a hot needle, as Roux did, Driesch shook seawater containing two-celled embryos so that the blastomeres were separated from their partners, but none were killed. Driesch then allowed the isolated blastomeres to develop and found, to his astonishment, that each produced a normal,

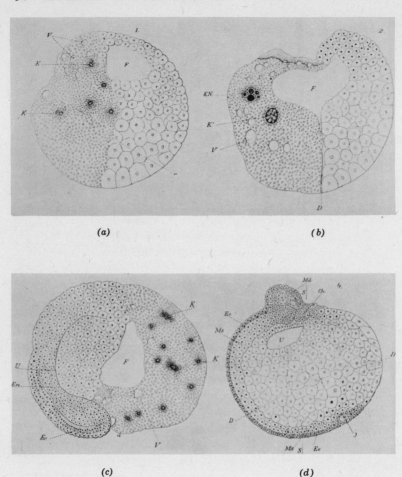

(a)   (b)

(c)   (d)

Figure 2-4   (*From Willier and Oppenheimer,* Foundations of Experimental Embryology, *1964*).

although somewhat small larva, called in the sea urchin a Pluteus (see Figure 2.5). Driesch's results were in direct opposition to the prediction of the mosaic theory. Instead of seeing differentiation as a result of parceling out of hereditary units, Driesch concluded that the embryo was a series of cells bound together as a self-adjusting whole, what he called an "harmonious equipotential system" (i.e., where all parts are equivalent in their potential for producing a whole new organism). If each blastomere, even when separated from other cells in the embryo, could still develop into a full adult,

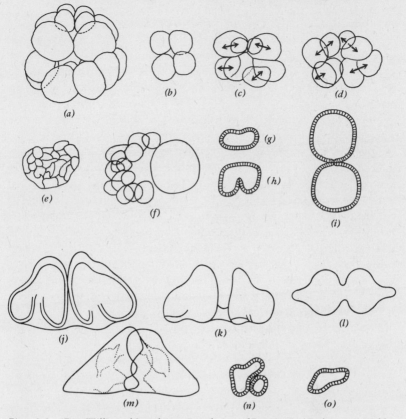

Figure 2-5   (From Willier and Oppenheimer, Foundations of Experimental Embryology, *1964.)*

then there must not have been any qualitative separating out of hereditary material as Roux had postulated.

Driesch observed, however, that if the two blastomeres were left together and the fate of their daughter cells traced, each blastomere would give rise to a different part of the embryo. In other words, when a two-celled stage is allowed to develop normally, each cell gives rise to qualitatively different sets of tissues in the adult organism. The fact that when the cells are separated each can produce a whole, complete embryo suggested to Driesch that cells had an inherent ability to adjust themselves to varying circumstances—that is, differentiation might be the result of cellular responses to both internal and external conditions. The "harmonious equipotential system" was cast in mathematical and physical terms and Driesch attempted to devise an explanation for his results in terms of external influences that might act on embryonic cells.

Ultimately, however, Driesch despaired of finding a causal-mechanical explanation for development and, by the early 1900s, had abandoned experimental biology almost completely for philosophy. He later became professor of philosophy, first at Strassburg, and later at Leipzig, advocating a neo-vitalism that represented a total reversal of the mechanistic views of his youth.

The apparent contradiction between the Roux and Driesch experiments was the result of differences both in the experimental techniques used and in the species of organism involved. Several years later other embryologists showed that if the one punctured blastomere in Roux's frog embryo were removed from physical contact with its partner, the latter would develop into a full-fledged, viable embryo. The physical contact between the injured and the normal blastomere in some way altered the developmental process. In addition, the different results seem also to have been to some extent the result of differences between the sea urchin and frog embryo. Biologists have always been aware of the fact that results that apply to one species may not generally be applicable to all species. In the case of Roux and Driesch, each erred by drawing general conclusions from work carried out on a single species. As tempting as such practice often is, all too frequently artificial dichotomies or generalities have been perpetuated by failure to recognize the importance of species individuality.

In a more profound sense the Roux-Driesch controversy involved the distinction between internal and external factors as the causal agents in differentiation. To Roux, who emphasized internal factors, the embryo contained within its biological makeup the elements necessary to guide its own development in an orderly and systematic way. Developmental steps were programmed in the sense that normal biological processes completely within the embryo controlled the sequence of events. Development was a process of self-determination. To Driesch, on the other hand, the biological processes of the embryo—particularly steps in differentiation—were strongly influenced by conditions external to the embryonic cells. The harmonious equipotential system that Driesch saw in the embryo received and responded to stimuli from the outside. The embryo was always adjusting to, and thus changing as a result of, environmental stimuli that impinged on it. Such factors as a cell's position within the embryonic mass, its degree of contact with the outside environment, the amount of contact with other cells, and the effects of gravity were, to Driesch, important causal agents in development. The regularity of developmental events was the result of the continual adjustment of the embryo to various external changes. Cells coming to lie internally always behaved in a certain way compared to cells that came to lie externally. The behavioral responses were translated into varying pathways of differentiation. It was therefore in the adjustment of the embryo to constantly changing conditions—changes caused both by its own growth as well as those imposed from the outside—that Driesch saw the key events of differentiation. Unlike Roux, Driesch saw development less as a series of rigidly programmed events than as the response of a living entity, the whole embryo, to varying life conditions. The mosaic theory was, to Driesch, too simple a mechanism for the complex series of events that he assumed must underlie developmental processes.

Neither Roux nor Driesch espoused internal or external causes exclusively. The difference was partly, at least, one of emphasis. Yet Driesch's approach led him ultimately to reject the notion that the causes of differentiation could ever be understood in rational, physico-chemical terms. The constant adjustability of the embryo defied reduction to material causes; it was, in Driesch's final view, the result of a vital force that no amount of experimentation could analyze or describe. On the other hand, Roux found his own conclusions compatible with continued emphasis on the physicochemical, experimental approach to embryology. His work continued in this direction through the first several decades of the twentieth century, although his most innovative period was between 1888 and 1900.

Roux's Program for Entwicklungsmechanik

During the later nineteenth and early twentieth centuries, Roux became the chief apologist and most vociferous advocate of the experimental and mechanistic approach to biology. From the 1890s until his death he devoted enormous energy to expounding his program for research in embryology. Given the impressive name *Entwicklungsmechanik*, Roux's method sought to find, through physical and chemical means, an explanation for how developmental changes are brought about. *Entwicklungsmechanik* is *roughly* translated as "developmental mechanics;" Roux himself took considerable care in selecting the name for his research program, rejecting such terms as *Entwicklungsphysiologie* because they did not adequately reflect the strong mechanistic bias that he wished to give to the new school of thought.

To provide a forum for experimental work in the *Entwicklungsmechanik* tradition, as well as to publicize his views, Roux founded a new journal, the *Archiv für Entwicklungsmechanik*, in 1894. In the introduction to Volume I, Roux explained what his new science embodied.

"Developmental mechanics or causal morphology of organisms, to the service of which these 'Archives' are devoted, is the doctrine of the causes of organic forms, and hence the doctrine of the causes of the origin, maintenance, and involution [Rückbildung] of these forms.

"Internal and external form represent the most essential attribute of the organism insofar as form conditions the special manifestations of life, to which the genesis of this form itself in turn appertains."

The key would is *causal*. Roux was interested in unraveling the causes behind changes during development. The terms cause or causal have, of course, many different philosophical meanings. Unlike Haeckel, however, cause meant to Roux the physical and chemical antecedents that resulted in particular embryonic structures instead of phylogenetic history. To Roux, causal meant material cause—that is, molecular and mechanical interactions.

The mechanistic aspect of Roux's program was emphasized when he repeatedly pointed out that biological research should seek the same kind of explanation as the physical sciences.

"Since, moreover, physics and chemistry reduce all phenomena, even those which appear to be most diverse, *e.g.*, magnetic, electrical, optical, and chemical phenomena, to movements of parts, or attempt such a reduction, the older more restrictive concept of mechanics in the physicist's sense as the causal doctrine of the movement of masses, has been extended to coincide with the philosophical concept of mechanism, comprising as it does all causally conditioned phenomena, so that the words 'developmental mechanics' agree with the more recent concepts of physics and chemistry, and may be taken to designate the doctrine of all formative phenomena."

Thus Roux tried to resolve the complex problem of differentiation, with its myriad component parts, into mechanical causes such as the qualitative separation of determinants by cleavage. The important features of Roux's analytic procedure was that it led to the formulation of experimental tests. *Entwicklungsmechanik* introduced in a highly formal way the experimental method into embryology. In a very real sense he created the field of experimental embryology by separating study of the causes of development from the more complex field of general morphology.

The Reaction to the Entwicklungsmechanik Program

Roux's method gained almost immediate support, especially among younger embryologists in the United States and Europe. In the early 1890's it was taken up by T. H. Morgan (1866-1945), E. G. Conklin (1863-1952), Ross G. Harrison (1870-1959), and E. B. Wilson (1856-1939), all orginally trained as embryologists in the classic morphological tradition. Roux's approach appeared as a fresh new wave on the dull and placid ocean of descriptive morphology, and it caught the imagination of a whole new generation of workers. And, when the Marine Biological Laboratory at Woods Hole, Massachusetts inaugurated its famous and fashionable Friday evening lecture series for 1894 with a translation of Roux's Manifesto (from Volume I of the *Archiv*), excitment became intense. C. O. Whitman, then director of the Laboratory, advocated the development of "biological physiology," using as his chief examples, among others, Roux's experiments on the frog. Not surprisingly, he saw its experimental methodology as the most crucial feature of the new approach. Whitman tried to urge on his colleagues in the United States these methods that he saw gaining ground in Europe. To do so, he maintained, would not only advance biology along new and more fruitful lines, but it also would fulfill the highest aims of the classical morphologists, who always sought to see life as whole, that is, form and function together. "The association of morphological and physiological research enlarges the field of vision on both sides," he wrote, and "converts half-views into whole views." It is interesting that many of those workers who quickly embraced *Entwicklungmechanik* in the United States were young (Whitman, the oldest, was 49; Morgan was 26, Conklin was 29, and Wilson was 36) and associated with the Marine Biological Laboratory at Woods Hole. This organization, founded in the late 1870s by a group of young workers led by Alphaeus Hyatt, was modeled on the Zoological Laboratory at Naples and was intended as a means of fostering serious (that meant, at the time, European-style) research, particularly along physiological lines (see Figure 2.6). It was at the Marine Biological Laboratory that experimental embryology of the *Entwicklungsmechanik* school was first fostered seriously in the United States.

Those who became *Entwicklungsmechaniker* sought in various physical and chemical agents the causes of cell differentiation—their major goal. They subjected embryos of many species (mostly marine invertebrates because they were easy to obtain and culture in the laboratory) to a wide variety of factors. Embryos were pressed between glass plates (Driesch), centrifuged (Roux, Loeb, and Morgan), placed in calcium or magnesium-free seawater (Herbst), and tied in two at the egg stage (the young Hans Spemann). Sometimes these operations resulted in normal development, sometimes in abnormal development. The main purpose of such work was to try and distinguish between Roux's hypothesis that factors for differentiation were internally determined and Driesch's idea that they were caused by external

Figure 2-6

factors. It is important to realize that this question has not yet been resolved, although the weight of modern embryological evidence suggests that differentiation is genetically determined. However, the process of triggering certain genes at certain times during development may be external to the cell being triggered, or it may be wholly an internal matter.

There were others, of course, who spoke out against or simply ignored the new approach. Old-style morphologists felt that experimental trickery with the embryo produced results that had no counterpart in nature. Under natural conditions, for example, embryos seldom have to undergo centrifugation, mechanical pressures on the egg, ion-free seawater, or destruction of one blastomere. What could such experiments tell about normal processes of development? they asked. Haeckel, for example, never mentioned Roux's work and never showed any signs of altering his own views on reproduction and growth. Yet most of the opponents were, in fact, older workers whose denunciations did not deter younger enthusiasts. Despite Roux's vociferous publicity, *Entwicklungsmechanik* had the advantage of proving its own merits and thus gaining ground in the face of some opposition from the morphological establishment.

Although the *Entwicklungsmechanik* school developed into a widespread and innovative branch of biology, Roux himself became more and more rigid in his thinking and his views and less open to new ideas and evidence as the years went on. He never substantially changed his views of development, despite the growing body of evidence the embryos did have the ability to adjust to external factors in a variety of ways. Roux's advocacy of experimentation tended to mean his *own* experiments. He did little in the way of any original thinking in the period after the mid-1890s and presented no new experimental ideas. His initial ideas were perpetuated because they fell on fruitful ground and occupied the attention of workers who were tired of morphological debates. As E. B. Wilson wrote in the mid-1890s, experimental embryology had come to dominate the biological scene as Darwinian theory had done in the previous generation.

"A remarkable awakening of interest and change of opinion has of late taken place among working embryologists in regard to the cleavage of the ovum. It is perhaps not too much to say that at the present day the questions raised by these experimental researches on cleavage stand foremost in the arena of biological discussion, and have for the time being thrown into the background many problems which were but yesterday generally regarded as the burning question of the time."

Perhaps the single most important contribution of the *Entwick-lungsmechanik* school is the method of analysis that Roux and especially Driesch developed. It was not the conducting of experiments *per se* that was so critical; after all, simply poking an organism to see what it will do or cutting off embryonic regions to see how development proceeds are also experiments. The importance of the Roux and Driesch experiments was that they embodied the test of a specific hypothesis in a rigorous, either-or fashion. As pointed out earlier, Roux's hypothesis of mosaic development led to a specific prediction (qualitative division of hereditary units during cleavage) that could be tested by a specific experiment. No matter which way the results of that experiment came out—whether development was normal or abnormal—the data spoke directly to the initial hypothesis. If development was normal, the mosaic theory could be discarded. If development was abnormal, the mosaic theory was strengthened.

In contrast, the more random trial-and-error approach to experimentation, the "poke and see what happens" method, could provide only a minimal amount of information: the results could not help in the verification or rejection of more general ideas, because no appropriate question had been formulated. It was thus in the method of formulating questions—as testable hypotheses—that Roux, Driesch, and the early pioneers of *Entwick-lungsmechanik* made their most profound contribution. This method was to spread rapidly through the previously descriptive areas of biology during the early twentieth century. The following chapters will describe the movement of the new experimentalism and its physico-chemical bias and the ultimate metamorphosis of that bias in the areas of heredity, general physiology, and evolutionary theory.

The Roux-Driesch controversy illustrates a phenomenon encountered frequently in the history of all sciences, biology included: the conflict between two competing progessional "schools of thought." Roux and Driesch operated from separate academic centers during the 1880s and 1890s and thus influenced initially different populations of students. Driesch worked for most of his experimental career at the Naples Station. Here he had an enormous influence on the many experimental biologists who passed through this international center. Morgan, for example, was greatly influenced by Driesch through their association at Naples. Roux, on the other hand, was associated with university positions throughout his career (Leipzig, Breslau, Innsbruck, and Halle). He was director of several institutes during his career; in the German university system of the time institutes were often plums—research groups over which the leader had almost total control—and their directors often soared to fame through the work of the "team." In some cases, one institute was created in order to challenge and (hopefully) supplant in influence another. Roux was not easy to work with, however, and exercised such rigid and unbending control that he produced few students who continued to do fruitful work into the twentieth century. Roux's influence was spread far more widely by the journal that he founded and edited. By exerting dominant control over the *Archiv* for 30 years, Roux made certain that the tenets of *Entwicklungsmechanik* in general, and his mosaic theory in particular, were well disseminated throughout the world biological community.

The result of such conflicts, as illustrated by Roux and Driesch, is that opposing views are often hardened into rigid and extreme positions that ultimately do not conform well with reality. Yet, as it frequently happens in the academic world, prestige and advancement rest on linking one's name to an idea, however wrong that idea may eventually turn out to be. While many scientists do change their minds, many others do not. Roux was such a case in point. He never admitted the mosaic theory had limitations, despite the varied evidence against it. De Vries was similar. Both men died retaining the same general ideas they had originated 30 years previously. The result was that neither Roux nor De Vries contributed as much as they otherwise might have to an understanding of the problem they investigated. With Driesch, on the other hand, the position he took in response to Roux was less rigid but equally as extreme in another direction. Faced with the mosaic theory, he retreated to a nonmaterialistic stance that ultimately forced him to abandon biology altogether. Like Roux, Driesch was less productive than he otherwise might have been had he not fallen victim to the formulation of biological theories in terms of a single, oversimplifying, and rigid school of thought. The problem with such competing "schools" is not that they have a viewpoint but, instead, that the viewpoint of one is often molded more in reaction to the other than out of careful study of the biological phenomenon at hand.

CHAPTER III

Revolt from Morphology II: Heredity and Evolution

I N THE FIELD of heredity the revolt from morphology was intimately connected with the development of Mendelian genetics after its rediscovery in 1900. In the later nineteenth century theories of heredity had been developed by biologists of enormous reputation and influence, from Darwin and Ernst Haeckel to Hugo deVries and August Weismann. A major reason for increasing interest in heredity was the gap made apparent by Darwin's theory of natural selection. Because the theory of natural selection lacked a workable concept of heredity, most biologists had become keenly aware of the need to understand how variations originated and were passed on to succeeding generations. As we saw in Chapter II, the fact that no really acceptable theory became known to the biological community at large was one reason for a widespread discontent with the Darwinian theory.

The various theories of heredity published between 1859 and 1900 differed in their terminology and often in the basic mechanisms they proposed. But they also had several basic characteristics in common. One was that they all proposed a particulate concept of heredity, the idea that hereditary information was stored and transmitted in particles residing in the germ plasm. At fertilization, particles from the male parent, carried in the sperm, are joined with particles from the female parent, carried in the egg. The ways in which the particles interact to form the heredity of the offspring varied from one theory to the next. But most workers had come to see that the older concepts of "blood heredity," or "blending" inheritance, were inadequate to explain the increasing number of facts about heredity in different animals and plants. Some sort of particles, which preserved intact their hereditary information, were seen as necessary to account for such phenomena as (using modern terminology) complete dominance, mosaicism, reversion to ancestral traits, or the appearance of extra digits (as in polydactyly).

Another common feature of all theories of inheritance in the later nineteenth century was their attempt to relate the process of hereditary transmission to other biological problems, such as embryonic differentiation, cell physiology, and evolution. The most ambitious attempt along these lines, and peculiarly representative of much nineteenth-century biological thought, was the work of August Weismann, discussed previously in Chapter I. At the time, a theory that did not attempt to account for such additional problems was thought to be incomplete. It is a testament to the wide-ranging concerns of nineteenth-century biologists that they saw the integral connection between physiology, cytology, heredity, and development. It was also the downfall of nineteenth-century biologists to have tried to attack problems that were too large and too complex to be investigated by the available concepts and techniques.

The rediscovery in 1900 of Mendel's 1866 paper provided an important source of new ideas about the hereditary process. At first Mendel's work was not wholeheartedly accepted by all members of the biological community. Ultimately, however, it came to be regarded as the best means of bringing together results of breeding experiments with the accumulated facts of cytology. The revolt against morphology was expressed in the field of heredity through the establishment of the Mendelian theory of heredity between 1900 and 1920.

Before considering the reception of Mendel's work after 1900, it is necessary to understand to some extent the ongoing controversy toward the end of the nineteenth century between proponents of continuous and discontinuous heredity.

Continuous and Discontinuous Variations

By 1890 an important issue had been raised for all biologists concerned with heredity and evolution. This question was: is inheritance continuous or discontinuous, and on which type of inheritance does natural selection operate? The issue arose directly out of Darwin's work and grew into a controversy of major proportions. Basically, the term "discontinuous variation" referred to the variations that occurred in recognizable, discrete forms with no gradations between one and the next. For example, proponents of this view of heredity would maintain that human eye color, or flower color, could be either one or the other of several possible colors. On the other hand, the term "continuous variation" referred to the variations that demonstrated a graded series, running from one extreme of the character to the other. To proponents of continuous variation, between the condition of brown and blue eyes, there might be a whole spectrum of intermediate colors. The question that many biologists were concerned in answering around the turn of the century was: which type of variation is actually inherited, as opposed to being influenced by the environment, and thus on which natural selection can act? If one or the other type could be shown to be not inherited, then that type could be dismissed as of little or no evolutionary importance. It was largely in context of its relationship to the mechanism of evolution that the argument over continuous versus discontinuous variation was placed.

In the *Origin of Species* Darwin had claimed that selection acts primarily on small individual variations of a more or less continuous sort. Though he could not prove it, he assumed that such variations were, for the most part, inherited. Among biometricians, some studied variation on a population level by statistical means. Led first by Darwin's cousin, Francis Galton (1822-1911) and later by Galton's student Karl Pearson (1857-1936), the biometricians tried to develop an understanding of hereditary patterns by quantitative studies of characters within a population.

The basic methods of the biometricians in general were measurement of specific visible characteristics of organisms and statistical analysis of the collected data. Because they studied characteristics that, in most populations, are determined by the interaction of heredity and environment, the data that biometricians collected usually showed "normal" distributions (see Figure 3. 1). This led many biometricians (Galton was less sure on this point) to conclude that most variation is of the continuous sort, in agreement with Darwin's qualitative judgment. Thus the biometricians rejected the idea of discontinuous inheritance, either as a general phenomenon, or as having any relation to evolution.

One example of Galton's generalizations, the "Law of Filial Regression," will give some idea of the kinds of hereditary theories espoused by the biometricians and typical of the period before the rediscovery of Mendel. Galton's Law of Filial Regression was basically a concept of blending inheritance based on the idea that the offspring of every generation were more like the mean of the population as a whole than like the mean of the two parents (unless that mean also corresponded to that of the whole population). For example, the offspring of a very tall and moderately short parent would be less likely to be intermediate between the two parents than closer to the mean of the short parent, because the latter would be closer to the mean of the population as a whole. As Galton wrote in *Natural Inheritance*, "the stature of adult offspring must on the whole be more mediocre than their

1	0	0	1	5	7	7	22	25	26	27	17	11	17	4	4	1
4:10	4:11	5.0	5:1	5:2	5:3	5:4	5:5	5:6	5:7	5:8	5:9	5:10	5:11	6:0	6:1	6:2

(a)

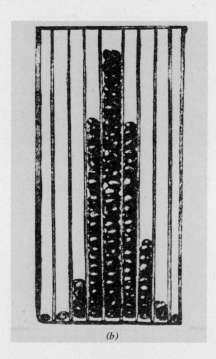

(b)

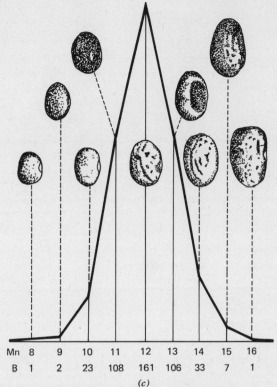

Mn	8	9	10	11	12	13	14	15	16
B	1	2	23	108	161	106	33	7	1

(c)

Figure 3-1   From Thomas Hunt Morgan, A Critique of the Theory of Evolution (copyright 1916, 1944 by Princeton University Press). Reprinted by permission of Princeton University Press.

parents, that is to say more near the mean or mid of the general population." Galton, like Darwin, held to a kind of blending theory of heredity, and consequently believed that new variations in a population, no matter how pronounced at first, would be weakened in effect by each generation of crossbreeding. Studying many human traits, such as stature, intelligence, weight, and body form, Galton tried to obtain quantitative data on their distribution and fate in a population. Although not a sophisticated mathematician, Galton did try to develop some statistical techniques for analyzing data on a large scale.

In 1900 Galton and his student Karl Pearson (a much more sophisticated mathematician) founded the journal *Biometrika*, which gave official status to the biometrical movement and provided an outlet for publishing results in this new line of inquiry. Although the biometricians published papers on many subjects, including statistical techniques, human inheritance, and eugenics, the single largest theme dominating their work remained the question of continuity versus discontinuity, and its relation to Darwinian selection. By and large, the biometricians were strict Darwinians and supported with their data the idea that evolution occurs through selection acting on myriads of minute, almost imperceptible (hereditary) differences among organisms of a population.

There were, however, difficulties that the theory of continuous variation encountered. The first and most obvious was that no clear-cut evidence existed to indicate that such variations were indeed inherited. A second was that examination of populations in nature showed that from one locality to another, variations often seemed to be discontinuous. That is, while within any one population the range of variations was continuous (forming a bell curve), between populations even of the same species, from different localities, the differences were distinct and pronounced.

These observations had given rise in the 1880s and 1890s to the idea that inherited variations were of the discontinuous sort while noninherited, environmentally induced variations were of the continuous sort. Many workers, therefore, came to believe that Darwin essentially was incorrect in his emphasis on the origin of species by selection of the very slight, continuous, variations among members of a population. Instead, it seemed that selection must act on some large-scale variations of the discontinuous sort. Those who believed in discontinuity rejected orthodox Darwinism as such. They by and large retained a belief in the efficacy of selection but looked to other quarters for an explanation of how variations originated.

Toward the end of the nineteenth century, the idea of discontinuity received its greatest elaboration in the work of William Bateson (1861-1926). In 1886 and 1887, Bateson had made a youthful and enthusiastic pilgrimage across the Asian Steppes to study the relationship between variation in the environment and variability within populations of organisms. In the Russian province Kazakstan, in particular, he found a very useful model for study. A number of lakes in this region have varying degrees of salinity, which can be arranged in a linear, almost perfectly continuous series of environmental circumstances. In making detailed studies of these lakes and their fauna, particularly a single species of shellfish distributed through most of the lakes, Bateson found that gradations in salinity were not accompanied by similar, proportional gradations in the range of specific characters of the organisms. In other words, the environment could be seen as representing a smooth, continuous gradation of physical characteristics, but variations among the organisms were nonetheless discontinuous. Bateson concluded that the variations giving rise to these differences among the shellfish must have been initially discontinuous and determined by heredity, not by the environment. These observations supported the view that the variations that were hereditary were discontinuous. The results of Bateson's systematic study of variability in many kinds of organisms were published in 1894 and titled *Materials For The Study of Variation*. Bateson related his work directly to the evolutionary question by concluding that selection can act only on discontinuous variations since they are the only ones that are inherited. Bateson's work drew sharp attack from the proponents of continuous variation and thus from the majority of biometricians and of course, neo-Darwinians, who saw only heresy in the idea of discontinuity.

By the turn of the century debates between proponents of continuous and discontinuous variation were rampant in the biological literature. Neither side was able to settle the debate with any satisfaction, because neither had a workable theory of heredity that agreed with the facts then available from breeding experiments. Followers of Darwin held more or less to a blending theory of inheritance or to some modified view of pangenesis. Followers of Bateson and other students of discontinuous variation rejected the idea of blending inheritance as well as Darwinian pangenesis.

Mendel's Works and Its Neglect

Mendel's original paper, "Versuche über Pflanzenhybriden," was first published in the *Proceedings of the Brünn Natural History Society* in 1866. Choosing the characters tallness and shortness in the pea plant (*Pisum sativum*) Mendel found curious results from his breeding experiments. If he crossed a tall plant with a tall plant, the offspring were all tall; similarly, a short plant crossed with a short plant produced offspring that were all short. However, when he crossed a tall plant with a short plant, the first group of offspring (termed by later workers the F_1 or first-filial generation) were all tall. However, when he crossed certain of these tall plants from the F_1 with each other, he repeatedly got a ratio among the second generation (the F_2) of three talls to one short plant. The short plants all bred true when crossed with other short plants. The shortness characteristic, although masked by the tallness character in the F_1 generation, reappeared unchanged in the next generation.

To explain these results, Mendel assumed that every parent organism contained two "factors" for each inherited characteristic (i.e., such as height, flower color, seed shape, or type of seed coat). One factor was inherited from the male parent, the other factor from the female parent. The organism might have both factors alike (i.e., both for tallness) or it might have the two factors unlike (such as one factor for tallness and one factor for shortness). It appeared to Mendel that some factors masked the expression of other factors when the two were together; the former he referred to as dominant, the latter as recessive. From these assumptions Mendel enunciated two generalizations, what nineteenth-century workers coined as "laws," of heredity (see Figure 3.2). The first was the Law of Segregation: that is, in the formation of germ cells, the two factors for any characteristic are always separated from each other and ended up in a different egg or sperm.

Mendel's second generalization, called later the Law of Independent Assortment, stated that the maternal and paternal factors for any set of characteristics segregate independently from those of any other set, each germ cell getting a random collection of factors derived from the male or the female parent. The assortment could be predicted statistically in terms of the laws of chance. Applying these laws to the analysis of seven characteristics in the pea plants he was breeding, Mendel was able to show the constancy and regularity of breeding results. In other words, his 1866 work showed the operation of what appeared to be regular and constant principles in heredity.

Despite the importance later attributed to Mendel's discoveries, this work was largely ignored for the 35 years. Only a few workers ever knew of Mendel's work, and Darwin, to whom Mendel sent a reprint of his paper, never read his copy (the pages of Darwin's reprint of Mendel are still uncut). There are many reasons for Mendel's neglect. It may have been because Mendel himself was unknown; or because his work was published in a relatively obscure journal which, although it circulated relatively widely, people may not have been in the habit of reading; or it may have been because the paper was highly mathematical, and seemingly inapplicable to living organisms.

Given the initial neglect of Mendel's work, it is all the more curious that his rediscovery was not the result of one but, instead, of three more or less simultaneous events. Hugo de Vries, Carl Correns (1864-1933), and Erich von Tschermak (1871-1962) were all engaged in studies of plant hybridization when, in 1900, they stumbled across Mendel's original paper. All appeared to recognize its importance in interpreting their own data and in deriving general laws about heredity. While evidence shows that probably only Correns had collected data of his own that might have led him independently to Mendel's conclusions, both de Vries and Tschermak were able to grasp the importance of the new laws when they found them explicitly laid out in Mendel's paper. Mendel's findings helped them interpret their own data better. Like Correns, de Vries & Tschermak were

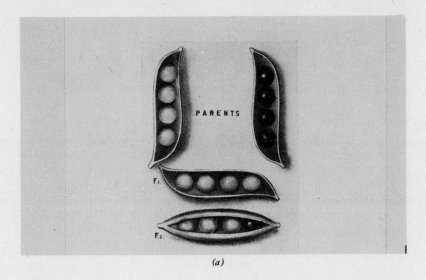

(a)

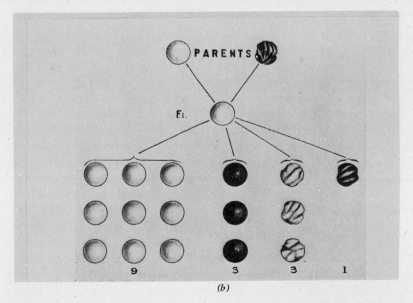

(b)

Figure 3-2   From Thomas Hunt Morgan, A Critique of the Theory of Evolution (copyright 1916, 1944 by Princeton University Press). Reprinted by permission of Princeton University Press.

prepared to recognize the significance of Mendel's earlier formulation as a result of the breeding experiments they had individually carried out. The simultaneity of events in this case refers more to the recognition of Mendel's importance than to the independent rediscovery of his laws.

Historians occasionally are prone to interpret occurrences of such simultaneity as illustrations of the adage that "the time is ripe" at certain points in history for certain ideas or events and is "unripe" at other times. What generalizations of this sort actually mean is difficult to say. In some way, "the time being ripe" means that workers were, in one era, more phychologically prepared to see and understand the relations of certain events or ideas than they were in some other era. The history of science is filled with examples of apparent simultaneity, and psychologists have known for a long time that what people actually see is dependent to a large extent on what they are prepared to see. In specific cases, however, the question confronting the historian is how to understand what the determining features were in an era when a discovery was accepted, or independently arrived at by several workers, and how these features are different from those of some other era in which the discovery, although technically or conceptually possible, was not made.

The Reception of Mendelism: 1900-1910

It is sometimes assumed that after 1900 biologists immediately accepted the Mendelian scheme and applied it to a variety of previously perplexing problems of heredity and evolution. However, there was a lag (between 1900 and 1910) during which many biologists voiced open hostility or skepticism about the new "laws" of heredity. There are many specific and general reasons for this initial skepticism within the biological community. An examination of several examples of early anti-Mendelism offers an understanding of the idealogical conflicts—especially concerning the areas of heredity and evolution—that carried over from the nineteenth century and characterized the transition period to the newer modes of of explanation characteristic of the twentieth century.

Despite the mathematical nature of Mendel's work, it was, ironically, the biometricians in England who were among the most vocal opponents of the new laws of heredity after 1900. Their opposition grew out of the longstanding debate over the role of continuous versus discontinuous variations in heredity. With its emphasis on "unit factors" the Mendelian theory was one of discontinuous variation. The biometricians, being largely neo-

Darwinians, were committed to emphasizing the importance of continuous variations; indeed, as we have seen earlier, their statistical analyses of bell curves and the like had strongly supported this prejudice. So influential was the biometrical school, especially in England, that its authority was in a position to eclipse the young Mendelian theory before it became established. The staunchest defender of the Mendelian scheme in England was William Bateson (1861-1926), whose work almost single-handedly raised it to prominence against the attacks of the biometricians.

Bateson first read Mendel's work on a train *en route* from Cambridge to London, where he was going to present a talk. He was so impressed that he immediately incorporated Mendel's idea into his lecture and, on returning to Cambridge, set about to have Mendel's paper translated into English. Furthermore, he began to test the Medelian theory on a variety of animals and plants. Preliminary results showed that in sweet peas, morning glories, and a variety of fowl, the Mendelian principles, or slight modifications, seemed to hold with remarkable regularity.

Bateson's initial espousal of Mendelism however, was immediately attacked by his former friend W. F. R. Weldon, an influential and outspoken biometrician. Weldon's arguments were based on the fact that Mendel's theory was just another example of a particulate theory of inheritance and hence emphasized discontinuity. As a biometrician and strong Darwinian, Weldon deemphasized discontinuous variations because they seemed to have little relation to evolution by natural selection. In 1902, as a result of Weldon's early attacks, Bateson issued a book titled *Mendel's Principles of Heredity: A Defense*, which answered the objections raised by the biometricians. The *Defense*, however, did not serve to convince biometricians who were already disposed to reject the Mendelian theory. Just after 1900, the battle of continuous *versus* discontinuous variations as being the best explanation of the hereditary process thus centered around the newly discovered Mendelian laws. The *Defense* polarized the position between Bateson and Weldon, and evolved quickly into a bitter and antagonistic series of polemics in the periodical literature as unproductive as that between Roux and Driesch had been.

The climax of this debate was reached at the 1904 meeting of the British Association for the Advancement of Science. On the platform at an afternoon session were both Bateson and Weldon, ready for a final confrontation over Mendelism. According to R.C. Punnett, who recorded his observations of the same some years later, Weldon spoke eloquently and with vehemence, but Bateson's arguments won the day. His data were more precise and, most important, his argument was clearly on the side of reason—of trying out the new theory before rejecting it out of hand. Bateson's victory was twofold. Not only had he won for Mendelian theory the right to be heard and considered as a viable account of heredity, but he had also won a major victory for the concept of discontinuous variation. Through his evidence and his ability to explain ideas clearly and forcefully, Bateson made discontinuity an almost implicit feature of any conception of heredity from that time on.

Bateson's generalship, however, was aided by the support of many others, especially a large group of practical animal and plant breeders. While visiting the United States to attend an agricultural conference in 1902, Bateson wrote back enthusiastically to his wife that everywhere he went he was greeted by agricultural experts waving copies of Mendel's paper in their hands. As he said, with evident satisfaction, "It is Mendel, Mendel all the way." It may perhaps seem curious that plant and animal breeders, with their primary concern for practical results, would have taken more readily as a group to Mendel's theoretical presentation than many of the more academic biologists.

However, Mendel dealt with the problem of hybridization and with laws that allowed them, within statistical limits to predict the kinds of offspring that should be expected from a given type of cross. Practical breeders were most familiar with ratios of offspring; although most breeders before 1900 had not looked critically or explicitly at them, these ratios were their daily business. Mendel's ideas gave some meaning and generality to the mass of breeding results with which practical breeders had so long been concerned. For the first time, breeders were able to make some sense out of their rule-of-thumb practices and to improve the efficiency of their breeding processes. Mendel told them, for example, that every organism showing a dominant trait was not necessarily a purebred (for that trait). Furthermore, Mendel's back cross (crossing an offspring showing a dominant trait with its recessive parent) provided a means of testing the genotype of any organism to find out whether it was a purebred or a hybrid. A testimony to the importance that the practical breeders saw in these theroretical studies was the establishment of the American Breeders' Association in 1903. Its purpose was to bring scientists and practical breeders together so that "each may get the point of view of the other and that each appreciate the problem of the other."

Despite Bateson's victory, the Mendelian theory still faced a strong group of critics in the years before 1910. Many of their objections grew out of specific questions of the general applicability of Mendelian laws or out of lack of concrete evidence concerning the physical reality of "factors." The early views of the American embryologist Thomas Hunt Morgan (1866-1945) illustrate well some of the major objections that many biologists had to Mendelism during these early years. Morgan's arguments are particularly interesting not only because they express some of the most frequently voiced criticisms of Mendel, but also because some years later (1933) Morgan was awarded the Nobel Prize in Medicine for having established the applicability of the Mendelian scheme as a general theory of heredity.

Among Morgan's many objections were the following:(1) Mendel's "laws" might apply to pea plants, but they were not demonstrable in a large variety of other organisms, especially animals; (2) the Mendelian theory of dominance and recessiveness would not account for the inheritance of sex in the one-to-one ratio that was observed (which sex factor, he asked, was dominant and which recessive?); (3) the Mendelian categories of "dominant" and "recessive" were not always as clear-cut as "tall" and "short" in pea plants; offspring often showed what seemed to be intermediate conditions between the supposed dominant and the supposed recessive character; (4) there was no proof of the existence of Mendel's postulated "factors," and the Mendelian scheme was thus a hypothetical construct that had no basis in reality.

If the Mendelian theory were merely a logical construct—a symbolic creation designed to agree with certain observations about crossbreeding— then it appeared to be ultimately no better, on methodological grounds, than the speculative theories of Weismann or Haeckel. In this view, Mendel's particulate theory was just another version of the same old tendency to postulate the existence of hereditary units that had no experimental or observational proof to supoort them. Even though the behavior of Mendel's particles might seem to "explain" more consistently some of the facts of breeding experiments, the scheme still represented a speculative approach to biology that many of the younger workers wished to avoid. Since no direct relationship betwen Mendel's "factors" and material bodies such as chromosomes had been proved in the decade between 1900 and 1910, it was difficult to know exactly where to draw the line the hypothetical nature of the Mendelian scheme. The tendency of enthusiastic Mendelians to add and subtract factors (i.e., to create multiple factor controls over certain characters) at any time in order to bring errant results into line underscored to many critics the basically speculative nature of Mendel's theory. As. T.H. Morgan wrote in 1909:

"In the modern interpretation of Mendelism, facts are being transformed into factors at a rapid rate. If one factor will not explain the facts, then two are invoked; if two prove insufficient, three will sometimes work out. The superior jugglery sometimes necessary to account for the results are often so excellently explained because the explanation was invented to explain them and then, presto! explain the facts by the very factors that we invented to account for them. I realize how valuable it has been to us to be able to marshall our results under a few simple assumptions, yet I cannot but fear that we are rapidly developing a sort of Mendelian ritual by which to explain the extraordinary facts of alternative inheritance. So long as we do not lose sight of the purely arbitrary and formal nature of our formulae, little harm will be done; and it is only fair to state that those who are doing the actual work of progress along Mendelian lines are aware of the hypothetical nature of the factor assumption."

Interestingly, in just a little over a year, Morgan had switched his position and become one of the most avid supporters of the Mendelian theory in America. He had in that period discovered the fruit fly, *Drosophila melanogaster*.

On the other hand, if the Mendelian "factors" were admitted to be real particles in the germ cells, then to many biologists—principally embryologists such as Morgan —the ominous spectre of preformationism was raised. Popular in the seventeenth and eighteenth centuries, preformation was the doctrine that the individual adult organism was already preformed in the unfertilized egg. Embryonic development was simply a matter of growth of the tiny, but otherwise perfectly formed adult, to a larger size. Since the early nineteenth century, embryologists had almost universally come to accept an opposing view, that of epigenesis, which stated that the embryo developed after fertilization out of formless organic material in the egg. Development was more than a quantitative change—it was also qualitative. To Morgan and others, the invocation of material particles as the bearers of hereditary information from one generation to another seemed to go back to the old idea of a preformed "character" in the germ plasm. Preformation suggested that the particle contained the character that was assigned to it, and that development was merely an unfolding of that character. Thus, to postulate particles was to put off the real question of how hereditary information from one generation actually determined the hereditary characters of the next. As an embryologist, Morgan was particularly sensitive to the preformationist elements in Mendelian theory. But there lies behind this somewhat catholic objection a consideration of even more significant proportion.

By postulating hereditary particles that were physically passed from one generation to the next, the Mendelian theory gave the impression of focusing attention on morphological units, as opposed to the processes by which these units functioned. An interest in structure without function was unthinkable to the younger generation, and Mendel's theory seemed at first glance to be primarily morphological. To many younger biologists, the really significant question were how do hereditary particles, if they exist, control the chemical processes of cells? why do some characters mask the appearance of others? how can such processes be studied physiologically? With an emphasis on processes, investigated by physiological means, young biologists saw the Mendelian scheme as a throwback to older theories dominated by hierarchies of morphological units whose function could not be investigated by any known means. Morgan made his point well in 1909 by quoting from a recent author who had written:

"The nature of present Mendelian interpretation and description inexplicably commits to the 'doctrine of particles' in the germ and elsewhere. It demands a 'morphological basis' in the germ for minutest phase (factor) of a definitive character. It is essentially a morphological conception with but a trace of a functional feature. . . With an eye seeing only *particles* and a speech only symbolizing them, there is no such thing as the study of *process* possible . . . It has been possible, I think, to show by means of what we know of the genesis of these color characters that the Mendelian description—of color inheritance at least—has strayed very wide of the facts; it has put factors in the germ cells that it is now quite certainly our privilege to remove; it has declared discontinuity where there is now evident epigenesis."

Morgan and many of his generation were no longer satisfied with explanations that had only a morphological component. Physiological considerations were crucial, if biology was ever to escape the deadening speculative theories characteristic of the later nineteenth century.

Lying more deeply than the objection to morphological units per se was a basic confusion that characterized much thinking early in the century. Preformed hereditary particles were called variously "factors," "genes," "unit characters," and the like. What many workers between 1900 and 1910 failed to grasp was the fundamental distinction between the herditary particle itself and the recognizable adult character to which it presumably gave rise. This distinction between what is called today the *genotype* and the *phenotype*, was explicitly pointed out in 1911 by the Danish botanist Wilhelm Johannsen, whose work on selecting pure lines I have already discussed in Chapter I. Johannsen emphasized that organisms did not inherit "characters" at fertilization, but only specific genetic components, or potentialities for those characters. The inherited potentiality he called the genotype. It represented (1) what the organism could pass on to the next generation, and (2) what the organism itself *could* (but not necessarily *did*) show as a visible adult character.

The distinction was crucial, since failure to make it led many workers—especially embryologists such as Morgan—into a quandry. Unconsciously, they saw phrases such as "inheritance of dark coat color" as implying that a sperm or egg cell carried the dark coat color condition. And such talk sounded too much like preformationism. Jmohannsen's distinction made it possible to think of hereditary particles not as fully developed adult characters, but as units guiding functional processes. If a "gene" (Johanssen introduced the term in 1909) was only the potential producer of an adult character, then the question of how that adult character was actually formed became a question open to study. Furthermore, Johannsen's conception made it possible to understand more clearly the Mendelian idea of dominance and recessiveness. Once the distinction became clear in the minds of those workers concerned with heredity and evolution, the easier it became to accept the Mendelian theory in a modern context.

Establishment of the Chromosome Theory: Mendelism 1910-1915

A persistent question since the rediscovery of Mendel's work has been whether Mendelian "factors" had any material basis in living cells. In 1902, Walter Sutton (1877-1916), a graduate student at Columbia University, had suggested the strong similarity between Mendel's hypothesis of segregation and the microscopically observable separation of homologous chromosomes during mitosis and meiosis. His suggestion had been confirmed by detailed cytological studies a year later, suggesting that Mendel's "factors" might be chromosomes, or parts of chromosomes. However, there were some immediate objections to this interpretation; most notably that there were far more characters expressed in the adult than there were actual chromosome pairs. Hence chromosomes must be groups of factors joined together. If this were true, then certain groups of characters should be observed as inherited together which, in these early years, was not the case. However, by 1906 Bateson had identified several "linkage groups" in plants, evidence that actually seemed to favor the chromosome hypothesis. These examples were not conclusive by themselves, but only suggestive. By the early years of this century a good deal was known about chromosome movements, both in mitosis and meiosis, but nothing was firmly understood about chromosomal function. Their involvement in the process of heredity was by no means clearly established between 1900 and 1910, although some work, most notably in E. B. Wilson's laboratory at Columbia between 1902 and 1905, pointed strongly in that direction. Similarly, the older work of Theodor Boveri (1862-1915) showed that embryos that were deficient in chromosomes developed abnormally, suggesting that each chromosome might carry the specific determiners for growth and development. By 1910 all these forms of data began to make it clear that chromosomes were cell structures that acted as the vehicles of heredity. Not all workers accepted this evidence as compelling, however.

Despite his recognition of linkage groups in several species of plants, Bateson firmly rejected the idea that Mendelian factors were parts of chromosomes. Like Morgan, Bateson had been trained as an embryologist—both had, in fact, studied under William Keith Brooks (1848-1908) at the Johns Hopkins University. Like Morgan, Bateson tended to see material theories of inheritance as bordering on the ancient doctrine of preformation. But there was more to it. Bateson had a personal antipathy to materialism in any form, an influence that he may well have imbibed from the idealist physicists who formed an influential circle at Cambridge between 1900 and 1920. Bateson's form of idealism ultimately prevented him from accepting the chromosome theory even after Morgan's group had shown that chromosomes could be mapped. The stubbornness with which Bateson held to his position placed him in the rear guard of the growing discipline of genetics. By 1920 he was an anachronism, staunchly claiming that there was no direct proof that "genes" (or Mendelian factors) had any material basis in cell structure. The lead passed, after 1910, to American workers, where a strong tradition of cell biology, fostered principally by E.B. Wilson (1856-1938) at Columbia University, was important in developing the conceptual relationship between breeding data and the facts of cell structure and function.

Among the American workers who most notably advanced the cause of Mendelism in the 1920s was the theory's former critic, T.H. Morgan. The change in Morgan's attitude was a result principally of his own work with the fruit fly *Drosophila melanogaster* (see Figure 3.3). In 1908 Morgan had begun breeding *Drosophila* in his laboratory at Columbia in order to determine whether large, de Vriesian mutation occurred in animals. Although no startling species-level mutations occurred in his cultures, Morgan found early in 1910 a curious white-eyed male fly in one of the breeding bottles. Although this new character, which had suddenly appeared, did not make the fly a new species, Morgan called the change a "mutation" and bred the mutant male to a normal (red-eyed) female. All the offspring showed the normal, red-eyed condition. However, when he crossed some members of this first generation with each other, he found that the white-eyed characteristic appeared again. Curiously, however, it always appeared in males, but never or only very seldom in females. If, on the other hand, a white-eyed male were mated with females from the first generation, half of the male and half of the female offspring had white eyes. Morgan found that theses curious results of eye color inheritance could be explained on the basis of the Mendelian theory—a fact that stimulated his rapid conversion from skeptic to enthusiast.

To account for the peculiar inheritance of white-eyes in relation to sex, Morgan further assumed that the Mendelian factor for eye color always segregated with the factor for sex (see Figure 3.4). Morgan was aware of the cytological work on sex determination that had been carried out in 1904 and 1905. Wilson and the cytologist Nettie M. Stevens (1861-1912) at Bryn Mawr had shown independently that the so-called "accessory chromosomes" (one pair of homologous chromosomes in which the partners are not structurally alike) were associated with the inheritance of sex. Wilson and Stevens had shown that in certain groups of animals and plants, inheritance of two normal-looking members of the accessory pair produced a female organism, while inheritance of one normal chromosome and one of the odd-shaped chromosomes produced a male (in today's terminology, the "normal" chromosome would be the X, and the odd-shaped, the Y). They had also shown that in other groups of organisms the reverse was true. In either case,

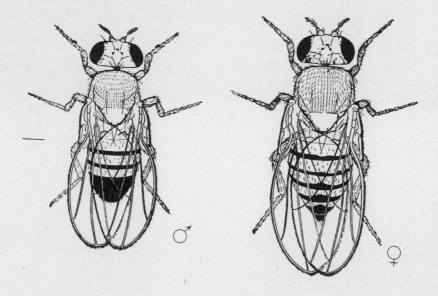

Figure 3-3   From Thomas Hunt Morgan, A Critique of the Theory of Evolution (copyright 1916, 1944 by Princeton University Press). Reprinted with permission by Princeton University Press.

however, the inheritance of sex had been shown to be related to the passage of specific chromosomes from parent to offspring.

Although in his early paper of 1910 Morgan had refrained from claiming that the "factor" for eye color was definitely linked to the accessory chromosome pair, he did try to explain his results by saying that both always segregated together. With this admission, however, it was not a long conceptual step to seeing that certain Mendelian characters could be physically linked to the accessory chromosomes. This was the beginning of a far-reaching theory of the physical basis of inheritance. Morgan referred to the white-eyed condition, and others thought to be determined by factors on the accessory chromosomes, as "sex-limited characters."[1]

[1] Today we refer to such traits as "sex-linked." Sex limited is a term that now refers to characters controlled by factors not lcoated on sex chromosomes, but expressed differently in males and in females (presumably different hormonal environment). Baldness is a sex-limited character, whereas hemophilia or red-green color-blindness is sex-linked. Throughout this chapter these terms will be used in their modern meaning.

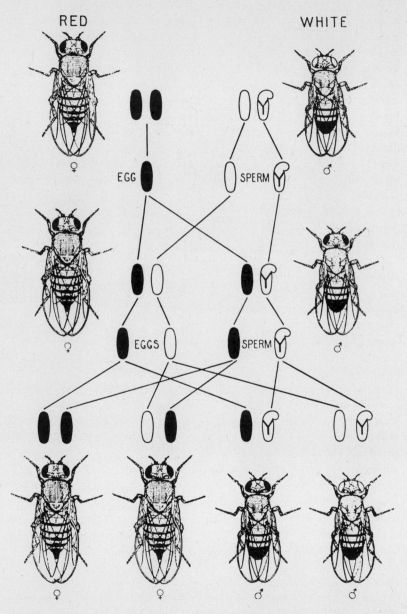

RED   WHITE

Figure 3-4 (From MENDELISM by Punet. By permission of Macmillan, London and Basingstoke).

With the discovery of sex-linked inheritance the way was paved for relating the Mendelian-chromosome theory to a comprehensive view of heredity. In the five years following the first *Drosophila* paper, astonishing results were obtained by Morgan and his co-workers. A means was developed for mapping chromosomes; the determination of sex was shown to be the result of a ratio of accessory to body chromosomes; specific chromosome markers were obtained and bred into pure lines for use in correlating breeding results with cytological studies of chromosome structure. The success of these endeavors was partly a result of the prodigious efforts by the "Morgan school" at Columbia. It was also because of the use of *Drosophila*, whose relatively short life cycle and easily satisfied dietary needs made it ideal for laboratory breeding. Bateson and the practical breeders had worked primarily with organisms whose generation times ran from three weeks to several months. *Drosophila* could produce a new generation every 10 to 14 days and thus contributed significantly to the rapid accumulation of data that Morgan and his group could analyze.

The development of scientific ideas is often greatly affected by the kind of physical or social conditions under which research is carried out. In addition to a favorable organism for study, Morgan also found a favorable environment for research in his department at Columbia University, and a favorable group of students to help him in his efforts. At the time *Drosophila* made its appearance on the Columbia campus, E. B. Wilson was chairman of the department of biology and a long-time friend of T.H. Morgan. Wilson was a man of dignified appearance and action whose monumental work on cytology, *The Cell in Development and Inheritance*, first published in 1896, showed his wide-ranging familiarity with cytological literature coupled with a strong interest in heredity and evolution. Sympathetic to Morgan and his interests in heredity, Wilson did everything possible to encourage the *Drosophila* research. Although the physical facilities in Columbia's Schermerhorn Hall were reasonably primitive, especially by modern standards, Morgan was able to convert a laboratory adjoining his office into what became known affectionately as the "fly room" and to equip it with some old tables and thousands of milk bottles for culturing *Drosophila*.

In 1909 and 1910, several eager and astute undergraduates at Columbia, having taken Wilson's introductory biology course a year or two previously, asked to work in Morgan's laboratory. These were A.H. Sturtevant (1891-1971), H.J. Muller (1890-1968), and slightly later C.B. Bridges (1889-1938). All of these men continued their graduate work under Morgan's direction and, along with others who also held positions in the fly room, helped to establish the major tenets of the chromosome theory of heredity.

In 1911 Morgan proposed the idea that if Mendelian factors were arranged in a linear fashion on chromosomes, there should be some way to map their relative distances apart. Knowing of an older theory of "chiasmatypes" proposed by F.A. Janssens (1863-1924) in 1909, Morgan suggested that breaks between chromosomes at synapses could result in an exchange of parts between two homologs. The farther two factors were apart on the chromosomes, the more frequently the exchanges should occur between them. Thus, by observing two characters at a time that were known to be contained on the same chromosome, the frequency of disjunction of the two characters in the offspring would indicate the frequency with which chromosomal exchanges had occurred. Morgan made his general proposition to Sturtevant as a problem for solution. Sturtevant, according to his own account, forgot his homework for a night and brought back to Morgan the next day the first map of a *Drosophila* chromosome. The map was crude, but it was a beginning and, within the next several years, a mapping procedure was worked out in great detail.

The group worked cohesively, but each individual had his own particular experiments and special interests. Morgan analyzed results and raised questions or larger problems to be attacked. Muller was especially adept at devising the precise experimental crosses that had to be made to test specific hypotheses that he or others proposed. He was also particularly interested in designing stocks of flies with specific chromosome markers, parts of chromosomes that could be observed visibly with the microscope and thus followed from one generation to the next. Aside from Morgan, Muller was the most inclined to think in broad, general terms. Bridges was the outstanding cytologist of the group. Not only his remarkable skill in making cytological preparations of chromosomes, but also his keen ability in spotting new mutants, or chromosomal aberrations, made him a key link in developing the relationship between chromosome maps and actual chromosome structures (see Figure 3.5). Sturtevant was particularly skilled with the mathematical relationships of mapping and often struggled most successfully with the quantitative arguments. The individual abilities complemented each other, providing a cohesiveness in those early years that allowed the work to progress with astounding rapidity. As Sturtevant described the working atmosphere:

"The group worked as a unit. Each carried on his own experiments, but each knew exactly what the others were doing, and each new result was freely discussed. There was little attention paid to priority or to the source of new ideas or new interpretations. What mattered was to get ahead with the work. There was much to be done; there were many new ideas to be tested, and

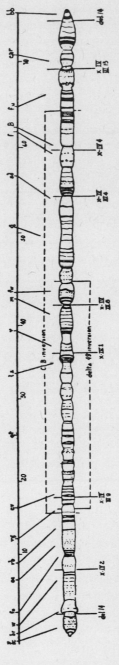

Figure 3-5 (From "A New Method for the Study of Chromosome Rearrangements and Plotting of Chromosome Maps," Painter, T. S., Science, Vol. 78, pp. 585-586, 22 December 1933. Reprinted by permission.)

many new experimental techniques to be developed. There can have been few times and places in scientific laboratories with such an atmosphere of excitement and with such a record of sustained enthusiasm."

Although intimately involved with the developing *Drosophila* genetics in the decades after 1910, neither Bridges, Sturtevant, nor especially Muller were mere appendages to Morgan; each was a well-established researcher. Bridges developed the comprehensive balance theory of sex, based on a chromosomal analysis of gynandromorphs and mosaics (flies that showed cominations of male and female charactertistics). Sturtevant developed the idea of position effect and most of the analytical techniques for mapping. And Muller, the most independent and intellectually aggressive of all, experimentally demonstrated the proportionality between dosage of X-radiation and frequency of gene mutation. For this work, done in the 1920s Muller was awarded the Nobel Prize for Medicine and Physiology in 1947. With a group of such young and remarkable associates, and with a cooperative organism such as *Drosophila*, it was little wonder that Morgan forged ahead at an enormous pace with his investigations.

By 1915 enough information had been accumulated by the *Drosophila* group for the publication of the epoch-making book, *The Mechanism of Mendelian Heredity* (by Morgan, Sturtevant, and Bridges). In this work, a solid basis was laid for the Mendelian theory both in its application to animal (and plant) inheritance as well as to the facts of cytology. In *The Mechanism of Mendelian Heredity* Morgan and his associates developed the idea that "factors," in Mendel's sense, were physical units, located at definite positions, or loci, on chromosomes. Each factor was viewed as a discrete unit physically separable from others around it by the process of chromosome breakage and recombination (chromosomes are known to break and exchange parts with the other member of each pair on occasion). *The Mechanism*, like Morgan's later *Physical Basis of Heredity* (1919) and *The Theory of the Gene* (1926), laid out the tenets of the Mendelian theory and tied breeding results to the facts of cytology. It was no longer possible to ignore the generality of Mendelian phenomena or to deny that Mendel's "factors" had some physical reality in the structure of the germ cells.

Chromosome maps were only one approach to establishing the physical reality of the gene. Later studies of deletions (where part of a chromosome becomes lost, with consequent loss of certain characters from the phenotype), inversions (where pieces of a chromosome are turned around), or duplication (where a sector of a chromosome is repeated, often affecting a phenotypic character), all contributed substantially to the idea that genes (or factors) were discrete units on chromosomes. When we realize that it took until the late 1960s (55 years after Morgan's work) for a single gene to be isolated—and at that from a virus—the deductions and achievements of the Columbia groups become all the more remarkable.

Reactions to the Mendelian Chromsome Theory: 1915-1930

The Mechanism of Mendelian Heredity (1915) met with almost immediate success, since the biological community had been slowly preparing for a shift in its general acceptance of Mendelism. And, of course, Mendelism was not new on the scene in 1915; however, the view of the theory that Morgan and his group provided was far more complete than Bateson or other earlier workers had produced. Morgan and his group gave the Mendelian theory a mechanistic-materialistic framework that attracted many of the younger biologist of the time (especially), who were imbued with the *Entwicklungsmechanik* spirit. A flood of papers, between 1910 and 1915 had emerged from the "Fly Room"— as the new results were broadcast widely through the biological literature. Morgan, ever a publicist, lectured on the new theories at Woods Hole, Columbia, Yale, many other universities, and at professional meetings. His considerable literary skills generated two additional books on Mendelism between 1915 and 1925, as well as numerous technical and semipopular accounts for the periodical literature. The new ideas were almost fully accepted by 1920 throughout the biological community.

Even after publication of *The Mechanism* and the many technical papers of Morgan and his colleagues, there were those who argued against one or another aspect of the Mendelian chromosome theory. For example, W.E. Castle (1867-1962) of Harvard University found that selection itself, pursued rigorously in one direction for numerous generations, seemed to alter the genes involved, to vary *in the direction of selection*. On these grounds Castle argued against the Morgan school's assertions regarding the stability of the gene (mutations were considered by Morgan and his group to be relatively rare events):

"Many students of genetics at present regard unit-characters as unchangeable. . . For several years I have been investigating this question, and the general conclusion at which I have arrived is this, that unit-characters are as modifiable as recombinable. Many Mendelians think otherwise, but this, I believe, because they have not studied the question closely enough."

Castle's assertions were later (1917-1919) shown to be untenable, since Morgan and his group could provide a better explanation of Castle's own results by the additional idea of "modifier genes." Modifier genes influence the degree of expression of a character; Castle's selection process had not modified the nature of the original gene, but only the number of modifiers acting on it.

One of the most difficult problems facing the new genetics was to account for how genes produced their effects. Although Morgan and his school wisely, in some respects, restricted their work to areas where they could perform actual experiments, that in itself did not serve to dismiss the question of gene function. The Morgan school studied gene structure—the physical arrangement and organization of genes on chromosomes—but there were those, such as Richard Goldschmidt, Director of the Kaiser-Wilhelm Institut für Biologie, who persistently tried to draw attention to the question of gene function. In a number of studies, culminating in the 1938 book *Physiological Genetics*, Goldschmidt attacked as too oversimplified the theory that genes were discrete parts of chromosomes. He maintained in 1940 that the theory of the corpuscular gene would, after 10 years, "be as dead as the dodo," precisely because it could not account for physiological and developmental processes. Goldschmidt preferred to see as the fundamental hereditary unit the entire chromosome, whose functioning was integrated into cellular metabolic processes. Goldschmidt was unsure what part of the chromosome was actually the genetic part, but he leaned toward the protein (chromosomes in higher plants and animals are composed of protein and nucleic acid) component, acting as an enzyme. Like Bateson, Goldschmidt was something of an idealist and objected, on philosophical grounds, to postulating the existence of gene particles that could not actually be seen. Nowhere in Morgan's or his group's work was it possible to do anything more than infer, from various lines of evidence, that discrete Mendelian genes actually existed. No one had ever then, or has to this day, seen an actual particle known to represent a single Mendelian gene. Goldschmidt was objecting partly to the faith people were placing in such dogma on inferential evidence. But he was also objecting to the fact that the Morgan theory was largely a structural theory, paying only lip service to functional questions.

Goldschmidt raised some important questions, and his attacks on the gene theory kept the question of gene function before the biological community for several decades. Substantively, however, he missed the mark. The most crucial feature of the work of the Morgan school was that at one and the same time they saw the intimate relationships between heredity, embryology, biochemistry, and evolution, and yet could essentially forget about the interconnections when asking question of *Drosophila*. The fact that Morgan could separate grandiose questions, which were not answerable by experiments, from simple questions, which were testable, allowed him and his group to push ahead in an area where at least some answers were readily obtainable. Controlled experimental conditions were seen by Morgan as providing the circumstances for detailed and quantitative analysis on limited, but answerable questions.

The new school of geneticists between 1915 and 1930 thus developed a strongly pragmatic and experimental approach to biology. Drawing analogies from the physical sciences, they extolled the value of experimentation and especially the importance of quantitative data. The Mendelian hypothesis was seen as having had one of its major influences by introducing experimental methods into a previously descriptive area of biology. Jacques Loeb (1858-1924), a close friend of Morgan's, and a champion of mechanistic biology (see Chapter IV) wrote in 1911 that new Mendelian science was "the most exact and rationalistic part of biology, where facts cannot only be predicted qualitatively but also quantitatively." Experimentalism triumphed in a clear and visible way with the acceptance of the chromosome theory. It is now no wonder that Morgan was irate when the Mendelian work was criticized for this very feature—for introducing unnatural elements such as artificial breeding, controlled environmental conditions, and the like into biology. To the younger generation of biologists these arguments were not only anachronistic, but they missed the point that the new direction in biology was finally able to take. As Morgan wrote in 1916:

"The objection has been raised in fact that in the breeding work with *Drosophila* we are dealing with artificial and unnatural conditions. It has been more than implied that results obtained from the breeding pen, the seed pan, the flower pot and the milk bottle do not apply to evolution in the "open," nature at large or to wild types. To be consistent, this same objection should be extended to the use of the spectroscope in the study of the evolution of the stars, to the use of the test tube and the balance by the chemist, of the galvanometer by the physicist. All these are unnatural instruments used to torture Nature's secrets from her. I venture to think that the real antithesis is not between unnatural and natural treatment of Nature, but rather between controlled or verifiable data on the one hand, and unrestrained generalization on the other."

The very basis of the Mendelian chromosome theory had been its quantitative and experimental approach. Through these methods, by limiting the questions asked and the factors that were varied, it had been possible to make out of the data from breeding experiments more sense than had been possible with the muddled and confused approach characteristic of nineteenth-century animal and plant breeders.

Equally important, the merging of cytology with the data from breeding experiments freed Mendelian genetics from the onus of being an old-type, speculative, nonmaterialistic theory. The merger gave genetics a physical reality that all previous theories of heredity had lacked. To Morgan and his group, the *Drosophila* work had set biology as a whole on a new plane by introducing experimental methods and by grounding the conclusions gained from breeding experiments in the firm reality of cell structure and function.

Mendelism and Evolution

Shortly after the work of the *Drosophila* group was well established through publication of *The Mechanism of Mendelian Inheritance*, T.H. Morgan was invited to deliver the Louis Clark Vanuxem Foundation Lectures at Princeton University. He chose as his subject the validity of the Darwinian theory as supported by evidence from modern genetics. It may seem surprising that Morgan chose this subject, especially since his own book, *Evolution and Adaptation* (1903), had been such a scathing attack on the concept of evolution by natural selection. However, Morgan was never one to stick stubbornly with old ideas once evidence accumulated against them. His lectures on the Darwinian Theory at Princeton thus reflect the contribution that the Mendelian chromosome theory made to Morgan's and others' understanding of the mechanism of natural selection. Historically, we are safe in generalizing from Morgan's reaction since, in his views on Darwin, he was representative of a number of biologists who, in the early years of the century, had voiced skepticism about natural selection.

The biggest stumbling block to a thorough understanding of Darwinian theory was lack of a consistent and workable theory of heredity. The *Drosophila* work had demonstrated to Morgan and others that genes for specific characters were inherited as discrete and constant units and were not altered from one generation to another by existing side by side in an organism with contrasting genes for the same character. For example, a white-eye gene could exist for one or more generations masked in effect by its red-eye counterpart (complete dominance) and still show up as an unaltered white-eye phenotype once it came together with a second white-eye gene. Recessive variations would not be "swamped"—that is, lost forever or diluted—even though they were not always expressed in every generation. Species could advance after all.

An equally important problem that the Mendelian-chromosome theory solved was that of the origin of variations. Morgan's own work, and that of his students (particularly Muller's on X-radiation), had shown that many types of genetic variability existed in a population, and that all could be the "raw material" on which selection acted. There was, of course, mutation—this time micromutation instead of the macromutations of de Vries—such as white eyes or vestigial wings. Mutations of this sort were discrete changes, affecting one or a few characters, which bred true to their new form once they occurred. Mutations were initially "mistakes" in replicating genetic units and were even found to occur at relatively constant rates for each mutation type (thus a new mutation for white eyes was found to appear about once in every 10,000 sperm or eggs produced).

There were sources of variability other than spontaneous mutation. One was chromosomal aberrations such as breakage and recombination, inversions, deletions, and duplications. Recombination forms new groupings of characters that might never have existed before in the same organism. Such new combinations could well have positive (or negative) selection value. Inversions and duplications were known to affect the expression of genes by changing their position in relation to other genes on the chromosome (what came to be called position effect). And, of course, a deletion could eliminate some genes from a family line, and/or allow the other members of the gene pair (recall that in higher organisms every trait is represented by two genes, one on each member of a chromosome pair) to be expressed. Furthermore, the discovery that there were more than two contrasting forms of some characters (e.g., there were many eye colors beside red and white in *Drosophila*), or that some genes could modify the quantitative expression of others (modifier genes as Castle had been working with) added great breadth to the form and kind of variations that could occur in natural populations. These different types of variability were what Darwin had described as the slight, individual differences on which selection acted. Morgan and his generation were able to study these variations experimentally and demonstrate their heritability in a way that had not been possible for Darwin or the neo-Darwinians at the turn of the century.

The genotype-phenotype distinction was of considerable importance at the time in helping Morgan and others to understand the mechanism of selection. It became clear that selection acted directly on the organism's phenotype—not on the genes per se. Thus the structural and functional characters that the organism displayed to the environment were those that selection favored or rejected. Two organisms, one of which carried a dominant and recessive gene and the other of which carried two dominant genes, would be subject to the same selection pressure because their phenotypes would be identical. However, in terms of the offspring they left behind, the two organisms would be different, qualitatively, and quantitatively. If, for example, the recessive gene produced an unfavorable phenotype, the organism bearing that gene would pass it on to one half of its offspring. The chances that the organism might mate with another organism in the population also carrying both the dominant and recessive gene would be determined by the frequency of the recessive gene in that population. If such a mating occurred there would be a greater chance that at least some (approximately 1 out of 4 according to Mendelian interpretations) of the organism's offspring would end up with two recessive genes and thus show the recessive phenotype. Overall, this organism's genotype would be less favorable, in terms of future generations, than the one carrying a double dose of the dominant gene (recessive genes, or mutations, are not always less favorable than the dominant or standard gene, however!). Morgan and others saw that populations of organisms could harbor much more variability than an inspection of the phenotypes might suggest. More important, this "hidden variability" could act as a store house for traits that might at some future time be more favorable under different environmental circumstances; it provided evolutionary "reserve," giving the population greater flexibility.

As the Mendelian-chromosome theory was worked out in its details, it became clear that it provided a mechanism of inheritance that was empirically verifiable. It was in agreement with Darwin's original assumptions that (1) at least some variability was in fact inherited, and (2) the vast bulk of such variability consisted of minute changes. But Morgan went even further. Seeing genes as discrete units allowed him to view evolution in terms of gene populations instead of only in terms of phylogenetic and taxonomic (morphological) similarity. Morgan saw looming in the distance the field that was to emerge a decade later as population genetics:

"The genetic analysis of a group of smaller species (i.e., varieties) would consist in finding out how the different genes are distributed amongst the members of this group. Phylogenetic relationship comes to have a differeent significance from the traditional relationship expressed in the descent theory; but this point of view is so novel that it has not yet received the recognition which we may expect that it will obtain in the future when relationship by common descent will be recognized as of minor importance as compared to a community of genes."

Morgan was not a statistical geneticist, and he did not customarily think of evolutionary problems in populational terms. But neither did a large number of those who in the first two decades of the century, were working much more directly on problems of evolution and the origin of species than Morgan. The full integration of Mendelian genetics with evolutionary thinking had to await the mathematical theory of natural selection of the 1930s. That theory was rigorous and quantitative and ultimately laid to rest all claims that natural selection was insufficient to account for the origin of new species. Prior to full synthesis, however, the Mendelian theory was able to fill in in a qualitative way the missing parts of Darwinian theory. Thus it allowed workers such as Morgan to put the full theory together in a way that had previously required an act of faith. Morgan was not alone in being unable or unwilling to grant Darwin's assumptions without experimental (or at least demonstrable) proof. Once this became available, between 1910 and 1925, he and others could move ahead to a more complete understanding of the process of evolution as laid out by Darwin 50 or 60 years earlier.

Conclusion

Although Mendelian theory itself was developed prior to the introduction of the *Entwicklingsmechanik* tradition, the new theory of heredity took hold only after 1900, when a generation of biologists existed who had been strongly influenced by the new experimental trend that Roux, Driesch and others had introduced. The tradition of quantitative, experimental analysis characteristic of *Entwicklungsmechanik* provided an atmosphere in which a new experimental science could gain foothold. Morgan paid his debt explicitly to the *Entwicklungsmechanik* tradition in 1898 when he wrote that Roux's work had for the first time brought the methods of physics and chemistry—particularly experimentation—into biology.

"The history of science teaches that by means of experiment chemistry and physics have made enormous progress; by means of experiment animal and plant physiology have become more exact, more profound studies than animal and plant morphology. . . Therefore, by means of experiment the student of the new embryology hopes to place the study of embryology on a more scienctific basis."

Entwicklungsmechanik had paved the way for greater acceptance of the Mendelian scheme by providing an atmosphere in which the new methods embodied in Mendel's work could thrive and expand.

The introduction of Mendel's theory into the biological community and the subsequent course of its history is a case study in the growth of scientific ideas. Despite its ultimate acceptance, Mendel's theory, like Darwinism, roused considerable skepticism within the biological community at its introduction. The common myth of so-called "scientific method" has it that scientists carefully and rationally consider all sides of an issue and come up with the "right" answer. Frequently, little is said of the points of view that later fail to win general support. The great figures of science are always pictured as those who were "correct," while the others, who were "wrong," are apologetically ignored.

With such a sterotype, scientists and others see new theories or ideas emerging like some sort of hidden treasure, the correct trail to which certain geniuses have uncovered while myriads of incorrect trails were being followed by lesser minds. The inception of Mendelian theory illustrates the misleading nature of oversimplistic views about scientific method. The general tenets of the Mendelian scheme are still accepted today. Yet, as we have seen, a man as eminent as T.H. Morgan was initially opposed to the new scheme. While Morgan's objections were to some extent peculiar to him, his general views were shared by many. These objections were, in many cases, valid; until enough examples of Mendlian ratios had been obtained in other organisms, for example, the argument that Mendel's laws might apply only to peas was valid.

Yet it is important to realize that there resides in the scientific community a certain conservatism, a certain reticence about immediately jumping behind any and every new idea. New ideas appear almost daily; consistent and well-integrated theories could not be build up over the course of time if old ideas were scrapped immediately on the advent of every new one. Discounting the conservatism of particular individuals, the scientific community as a whole is not as fast to change its shared beliefs as popular legend has it. It is not only churchmen or politicans who are reticent about accepting new ideas. Scientists, both as individuals and as a collective professional body, also share in a reluctance to abandon cherished ideas.

CHAPTER IV

Mechanistic Materialism and Its Metamorphosis: General Physiology, 1900-1930

IN 1911 THE German-born physiologist, Jacques Loeb (1858-1924), delivered an address titled "The Mechanistic Conception of Life" to the First International Congress of Monists,[1] convened by Ernst Haeckel in Hamburg. This address, published as a book under the same title in 1912, was Loeb's passionate testament to his belief that all aspects of living phenomena could be reduced, through laboratory analysis, to the fundamental laws of physics and chemistry. Loeb's own work on artifical parthenogenesis (stimulating an egg to begin development without fertilization) and animal tropisms had emboldened a mechanistic and reductivist bias that he had inherited directly from the apostles of the Berlin School of the 1850s. Loeb became for the the early twentieth century an eloquent spokesman for a starkly simple mechanist philosophy.

Loeb's views were warmly received by the assembled Monists in Hamburg and reached an astonishing vogue among biologists and nonbiologists alike by the 1920s. Loeb enunciated more clearly and explicitly than any of his contemporaries the logical extension of a methodology embodied in the *Entwicklungsmechanik* tradition. As Roux had hoped, biology was to become one with physics and chemistry with complex processes such as development, regeneration, and fertilization ultimately to be explained in atomic and molecular terms. To Loeb the time was almost present when this dream would be realized.

The mechanistic philosophy enunciated by Loeb was not, however, destined to rule the biological sciences for long in an unaltered state. Loeb's

---

[1] Monism is the general term for any type of philosophical belief in the ultimate unity of all knowledge. Both materialism and idealism are moniotic philosophies, as suggested by the fact that the Society of Monists was founded in 1906 jointly by Ernst Haeckel (materialist) and Wilhelm Ostwald (idealist).

simplistic models were too reminiscent of the Berlin reductionists of the previous century to settle easily in the minds of even the most rationalistic experimentalists, let alone the more descriptive biologists who held a natural aversion to mechanical models in living systems. What is more important, Loeb's single-minded attempt to reduce all biological phenomena to matter in motion was, by the 1930s, running counter to a new concern, appearing first in both physics and physiology from the early 1900s on. That concern centered around seeing the organizational interrelationships between parts instead of seeing parts in isolation. It involved new methods and criteria for studying complex phenomena without destroying organization by reducing all pheonmena to separate, component parts. This movement, visible to some extent in the work of Claude Bernard in the 1870s, occupied an important place in the studies of J. S. Haldane, L. J. Henderson, Charles Scott Sherrington, and Walter B. Cannon in the first three decades of this century. While not rejecting the idea that all physiological processes in-volved the interaction of atoms, these physiologists came to see that reducing all complex processes to molecular interactions missed a very important biological point. Living systems possess high degrees of *organization*—they are not simply random collections of molecules. Thus many biologists, while sharing Loeb's hope that all living phenomena could someday be understood in rational terms, ultimately rejected his approach. They fashioned from it what seemed to be a more satisfactory form of explanation that tried to account not only for the molecular mechanisms by which individual parts of a system functioned, but also for how those parts were *organized* (i.e., interrelated with each other to form the complex, living whole).

The present chapter will explore the development of mechanistic materialism during the early decades of the century, primarily through the work of Loeb, and its ultimate decline in favor of holistic materialism. I will try to relate the developments in mechanistic biology to develop-ments in other sciences such as physics, and in the broader areas of philosophy with the rise of the neo-Machian school.

The Origin of Loeb's Mechanistic Physiology

Jacques Loeb was a child of his time. Born in 1859, the same year as the publication of *The Origin of Species*, Loeb was ushered into the world at a time when long-cherished ideals were fast yielding to hard-nosed and pragmatic analysis in both the world of ideas and action. Political upheavals in the late 1840s had corresponded with the origin of a mechanistic physiology in

Berlin. Similar upheavals in the 1860s and 1870s throughout Europe were to have their parallels in Darwinian theory on the one hand, and the origin of *Entwicklungsmechanick* on the other. Although we have seen Darwin's work as that which the more analytical tradition of experimental embryology reacted against, Darwin's original work bears many of the same philosophical consequences as Roux's. Both ultimately saw living systems (individual organisms in their ontogeny or well-adapted species in their phylogeny) as the result of inanimate physical forces, operating continually throughout history. It was into this environment that Loeb was thrust and from which he fashioned a philosophy not only of biology, but of all society.

Having been greatly influenced by his youthful reading of Schopenhauer, Loeb became preoccupied with the question of freedom of the will—leading to an early interest in philosophical studies, which he pursued at the University of Berlin. Like Schopenhauer, Loeb concluded early in his life that the world was deterministic, and that freedom of the will was only a philosophical illusion. Unlike Schopenhauer, however, he was not pessimistic about the implications of determinism for the future of man. If determinism meant that human actions were based on discernible causes, as opposed to the noncausality implied by free will, then it should be possible to reorder society in more rational and predictable ways. The new society that he envisioned would be established on the basis of laws of human physiology and behavior that only a deterministic philosophy would uncover.

Loeb soon found that traditional academic philosophy did not offer the kinds of solutions to the problem of the will that he had earlier hoped it would. With great disgust at the verbal confusion and muddiness on which philosophical arguments seemed to thrive, Loeb decided that the best hope of understanding the nature of man's will lay in neurophysiology, not philosophy. So, in 1880, he entered the University of Strasbourg to study localization of brain function in the laboratory of Friedrich Leopold Goltz (1834-1902), a pupil of Helmholtz and strong proponent of the reductionist-mechanist viewpoint. After extensive work with Goltz, Loeb began to see that neurophysiological techniques, also, were too primitive to help elucidate the subtle and complex causes of animal, and especially human behavior. In 1886, therefore, he moved once again, this time to the University of Würzburg as an assistant to the great physiologist, Adolph Fick (1829-1901), a former student of Carl Ludwig (1816-1895), one of the Berlin medical materialists. This association proved to be a very fortunate one, not only because of what Fick personally could teach Loeb, but because he introduced the young assistant to his good friend, the plant physiologist,

Julius Sachs (1832-1897). Sachs, also the teacher of Hugo de Vries, was a great advocate of experimentalism and, along with Fick, a staunch mechanist. Loeb thus found himself in the bosom of a mechanistic tradition that was directly descended from the original reductionist school of Helmholtz and the Medical Materialists. At the time Loeb went to Würzburg, Sachs was interested in the study of plant tropisms, the response of plants to various external stimuli such as light, temperature, or gravity. The essence of tropisms according to Sachs was that they occurred as automatic, almost unavoidable, responses of the organism to a known set of conditions. In plants, Sachs had been able to show that tropistic responses often had distinct chemical or mechanical causes.

Profoundly influence by Sachs' work, Loeb wanted to raise the tropism concept, along with Fick's and Sachs' philosophy of mechanism, to a new life-style in science: determinism in physiological processes in general and animal behavior in particular, and the study of these processes by the tools of physics and chemistry. Emigrating to the United States in 1891, Loeb became one of the most vocal exponents of a new mechanism and its methodological counterpart, experimentalism. He carried into his view of every aspect of human life the mechanisitc philosophy that he adopted for biological problems. As a close friend and associate of many younger antimorphologists during the first decade of the twentieth century, he was enormously influential in advocating a new approach to old biological problems.

The mechanistic approach that Loeb had inbibed from Sachs was enhanced and given a specific direction by developments in chemistry toward the end of the nineteenth century. Of particular importance were those in the theory of chemical kinetics and electrolytic dissociation (the formation of ions in solution) during the 1880s and 1890s. At Würzburg, along with Loeb, was the great Swedish chemist, Svente Arrhenius (1859-1927), who showed in 1887 that compounds placed into solution dissociate into positively or negatively charged atoms (called ions) to a degree depending on the nature of the compound and of the solvent. Arrhenius, in turn, was in close contact with the physical chemist Jocobus Van't Hoff (1852-1911), the first Nobel laureate in chemistry (1901). In the 1870s and 1880s, Van't Hoff made a number of significant contributions to chemical theory: he worked out the dynamics of equilibrium reactions, developed explicitly the analogy between the kinetics of chemical reactions and the gas laws, proposed a physical theory to explain the osmotic pressures of solutions, and put forward the tetrahedral model of the carbon atom (the idea that a carbon atom bonds to four other atoms in such as way as to produce a molecule the shape of a regular tetrahedron). Both Arrhenius and Van't Hoff founded their chemical comcepts on the atomic-molecular theory and thus gave physical reality to the discreet particles with which they dealt (atoms, molecules, and ions). Loeb became a close friend of Arrhenius while still in Würzburg, corresponded with him regularly after coming to the United States, and served as his host when Arrhenius visited this country on various occasions. Through this contact, Loeb became not only versed in the new chemical and kinetic theories, but also enthused with the idea of applying these eminently quantifiable and precise concepts to the problems of biology. As he wrote of the atomic theory as it was understood in the 1890's: "Development of the atomic theory put science for a long time, and probably irrevocably, on a mechanistic basis. It marks perhaps the greatest epoch in the history of cognition."

In 1912 Loeb published his collection of essays titled, *The Mechanistic Conception of Life*. The essays covered two kinds of physico-chemical work: (1) studies of living cells, especially problems such as artificial parthenogenesis, fertilization, and osmotic regulation; and (2) studies on the causes of animal behavior, principally the nature of animal tropisms. To Loeb, the "mechanistic approach" meant the attempt "to analyze life from a purely physico-chemical viewpoint." This meant not only *reduction* of complex phenomena to the interaction of ultimate particles, but also implied that the methods of physics and chemistry were the only proper means of advancing knowledge about life. These methods included experimentation, with its emphasis on rigorous conditions, controls, quantitifcation, and the criterion of repeatability.

In 1899 Loeb found that he could cause an unfertilized sea urchin egg to undergo development simply by pricking it with a needle or changing the salt concentration of the seawater in which it was being cultured. In many cases Loeb was able to bring the artificially induced embryos to the larval stage or further. He also carried out similar experiments with frogs, raising some individuals from unfertilized egg to sexual maturity. Studies of the actual processes of fertilization itself showed that the sperm carried a molecular substance, which Loeb called lysin, thought to be responsible for activating the egg. Thus Loeb claimed that he could recreate the fertilization process in the laboratory without the necessity of the living sperm simply by supplying a physical or chemical agent that mimicked the molecular effect of lysin.

The experiments on artificial parthenogenesis created much excitement among biologists and laymen alike. To many biologists Loeb's work opened up a wholly new approach to understanding the molecular basis of life. If known chemical agents could recreate biological processes, then these pro-cesses must be purely molecular in origin. To laymen, the successful produc-tion of parthenogenetic organisms was both exciting and terrrifying, presag-ing the era in which life would be created in a test tube. In a less philosophi-cal vein, contemprary wits noted that ever since Loeb's discovery about salt water stimulating parthenogenesis, maiden ladies had expressed grave doubts about ocean bathing; one less subtle observer noted that weekends at the seashore did indeed seem to be remarkably fertile.

While Loeb's work on artificial parthenogenesis generated a considerable amount of interest, it was his studies on animal tropisms that most charac-terize his mechanistic approach. Loeb had noted that larvae of certain species of butterflies come out of the ground in the spring near the base of certain bushes where the eggs were laid the previous fall. As soon as they come from the nest, the larvae immediately crawl straight up the stalks of the bush and feed on the tender leafy shoots near the tip. In the laboratory Loeb found that the insects were positively phototropic; that is, if an insect were placed on a table in the dark, with a light source to one side, the larva would always orient itself toward the light. If the light source were suddenly moved to the other side, the insect would reorient itself. Loeb explained this phenomenon in terms of photochemical molecules in the insect's eyes. Light from one side stimulated photochemical substances in the eye on that side, but not in the eye on the other side. A chemical inbalance was thus created that when translated into a neural response, caused the organism to reorient itself so that equal quantities of light fell on both eyes. As an intriguing analogy, Loeb took particular delight in the selenium-eyed dog built about the same time by the brilliant design engineer, J. H. Hammond. Like the larval insects, the selenium-eyed dog oriented itself toward light by means of differential electric currents created when more light fell on one of two selenium discs (the eyes) placed at the "front" of the dog. A photochemical current was set up that activated a set of wheels on the illuminated side of the "dog," turning the machine toward the light. In true mechanistic fashion, Loeb saw the design of a machine with behavior identical to his caterpillers as direct support for the physico-chemical approach to the behavior problem.

Loeb went on to relate his tropism experiments to the adaptive behavior that insect larvae show in nature. When he brought branches of the bush on which the insects normally fed into the labroatory, but placed a light at the base instead of at the tip, the insects did not crawl up the stem toward the leafy shoots as usual. They turned themselves face downward and, because the leafy shoots were only at the top, starved to death. To Loeb, these insects had no free will and could make no decisions; they could only respond, mechanically, to a set of stimuli. They could not reevaluate a situation and adapt to the modified (and from a survival point of view, illogical) conditions imposed in the laboratory. The response was physico-chemical in origin but was beyond any willful control of the organism. The larvae were, in Loeb's words, "photochemical machines enslaved to the light."

From his concept of tropisms and from his view of organisms as chemical machines, Loeb formulated by the 1890s a completely deterministic view of society, morality, and ethics. He was back to the problem that had originally led him to physico-chemical biology: the nature of instincts and freedom of the will. To Loeb the source of all ethics was the instinct, the heritable foundation of behavior. No matter how subtle a behavior pattern might seem, Loeb assumed it could be reduced to a series of instinctual responses over which the organism had no direct control.

"We eat, drink, and reproduce not because mankind has reached an agreement that this is desirable, but because, machine-like, we are com-pelled to do so. We are active, because we are compelled to be so by processes in our central nervous system . . . . the mother loves and cares for her children, not because meta-physicians had the idea that this was desirable, but because the instinct of taking care of the young is inherited just as distinctly as the morphological characters of the female body. We seek and enjoy the fellowship of human beings because hereditary conditions compel us to do so. We struggle for justice and truth since we are instinctively equipped to see our fellow beings happy . . . . Not only is the mechanistic conception of life compatible with ethics: it seems the only conception of life which can lead to an understanding of the source of ethics."

According to Loeb's view, mores and customs should be developed from an understanding of the instinctual and deterministic bases of human action more than on the basis of tradition, superstition, and vague references to abstractions such as the "dignity of man." Human beings did not inherit a free will or specific tendencies toward abstract qualities, such as goodness or alturism. Instead, like phototropic insects, they inherited specific responses to specific stimuli, based on a determined set of molecular interactions. To Loeb, programmed behavior was adaptive—that is, it led to increased chances for survival of the species by contributing to the common good. But unless the conditions eliciting each programmed response were known, human beings, like the insects who starved because they could not turn away from the light, could find their own behavior patterns perverted by the artificial circumstances that social custom, tradition, and ignorance per-petuated as institutions of society.

Reaction to Loeb's Mechanistic Approach

Given the range of his theorizing and the extremity of many of his views, it is not difficult to understand why Loeb had as many critics as he had supporters. Most biologists agreed that his experimental work was precise and rigorous, and that the particular observations he made in a variety of fields were, as far as they went, factually correct. Where opinion divided was on the limits to which the mechanistic approach should be pushed in biology particularly and society in general. Yet critics and supporters alike knew that Loeb's view were serious and had to be dealt with honestly in their own rights.

First, there was the question of the generality of Loeb's observations throughout the biological realm. Could the specific phototopic response that Loeb had observed in a few species be extended to a general theory of *all* tropisms in *all* species, or even more, to a theory of all animal behavior? The protozoologists S. O. Mast (1871-1947) and H. S. Jennings (1868-1947), for example, claimed that Loeb's tropism theory had little bearing on understanding the behavior of unicellular organisms. Their own studies had suggested that trial and error was an equally if not more important factor in determining an organism's behavior than the rigid, stimulus-response mechanism that Loeb had postulated. Jennings and Mast argued that there was no evidence that a single kind of physiochemical mechanism was involved in the behavior patterns of all species; such a theory was to them a grandiose speculation not warranted by the facts.

Another criticism centered on the question of whether the mechanistic interpretation of pheonomena such as phototopism or fertilization, even if they were essentially correct, should be extended implicitly to the solution of problems in any or all other areas of biology. Behind this criticism lay a distrust of simple-minded interpretations, of "monistic" world schemes. To those predisposed to view his work critically, the mechanistc bias that Loeb held was little better than the grandiose speculations of a Weismann or a Haeckel. All were guilty, in this view, of generalizing a few facts into an entire system, and of choosing the data to fit a preconceived scheme. One German physiologist was vehement in his criticism of Loeb's tropism theories when he wrote in 1916 that Loeb was "the kind of investigator to whom the insignificant isolated facts are of no importance . . . . The glittering *theory* is everything to him, the facts only its servants. If the fact does not suit, then it will be forced into the theory."

A third criticism of Loeb's theory touched on the central issue of whether an exclusively mechanistic approach to biology was not, in the long run, an obstacle to new research methods and to new ways of viewing organic processes. Could it be, some asked, that the final solution to all biological problems would be to reduce them to a physico-chemical level? Was there never likely to be any more to it than this? The anatomist and student of biological form, E. S. Russel (1887-   ), put the question fairly and succinctly: the mechanistic approach did not "allow for the possibility that animals are more than mere machines and that neither their actions nor their perceptions can be satsifactorily analyzed in physiological, i.e., physico-chemical terms alone." To many, particularly a growing group of physiologists, a strict mechanistic interpretation limited instead of expanded the range of questions that could be asked about biological problems. There seemed no reason, therefore, to adopt the mechanistic method exclusively.

For those raised in the *Entwicklungsmechanik* tradition, however, Loeb's mechanistic biology was the very epitome of all they had been aiming to accomplish in their respective fields of embryology, cytology, and heredity. Loeb's approach was the logical and final answer to the Weismanns and the Haeckels whose speculative, metaphysical views had for too long dominated biological theory. In the introduction to a series of monographs for the J. B. Lippincott Company, Loeb, along with his coeditors T. H. Morgan and the physiologist W. J. V. Osterhout (1871-1939), explained the new mechanistic and experimental aims of biology for all the world to see.

"Biology, which not long ago was purely descriptive and speculative, has begun to adopt the methods of the exact sciences, recognizing that for permanent progress not only experiments are required but quantitative experiments. It will be the purpose of this series of monographs to emphasize and further as much as possible this development of Biology."

According to Loeb, Morgan, and Osterhout, the methods of exact science most directly available to the modern biologist were generally those of the physiologist. Hence experimental biology and general physiology "are one and the same science, in method as well as content, since both aim at explaining life from the physico-chemical constitution of living matter."

Loeb's mechanism had its own strong influence within the field of physiology proper. It was not limited to work with cells and biochemical systems alone. But as the mechanistic approach was applied to higher-level processes, such as the functioning of the nervous system, its limitations become more apparent to a broad spectrum of workers. The application of, and gradual disafffection with, the mechanistic philosophy as related to complex physiological processes can be seen in the development of ideas about function of the nervous system.

Reflex Arcs: Ivan Pavlov

Study of nerve conduction and the organization of the nervous system became one of the most important areas of neurophysiological research in the second half of the nineteenth century. Work on the nervous system grew out of two separate but equally (in their own way) influential traditions. One was the school of medical materialism, discussed in Chapter I—the Berlin school of Helmholtz. Their reductivist approach had concentrated on studying nerve conduction and the properties of isolated nerves. From this work had come the notion that nerves conduct at specific velocities, and that the velocity depends on the type of nerve being studied. Also from the Helmholtz school came the concept of threshold and the "all-or-nothing" response pattern in nerves. The latter maintains that a nerve cell will respond either completely or not at all to any given stimulus. A weak stimulus produces no response, a stronger stimulus a full response. The threshold represents the stimulus level that is just enough to stimulate the nerve; it is the turning point where a slight increase in stimulus strength causes the nerve to fire. By using electric stimulation, by altering temperature, ionic concentration of fluid surrounding the nerve, and by various other physical and chemical manipulations, the Berlin reductionists learned a great deal about the functioning of individual nerve cells and, more generally, about the bundles of cells making up an individual nerve.

An equally important tradition was that represented by the French and English schools, led principally by Charles Bell (1774-1842), Pierre Flourens (1794-1867), Claude Bernard, Pierre Magendie, and Michael Foster (1836-1907). The first three made important contributions to an understanding of the organization of the nervous system; the latter was a great teacher who established the principle of laboratory teaching in general and neurophysiology in particular in England in the 1870s. Bell and Magendie were the first to work out the distinction between afferent (leading toward the spinal cord) and efferent (leading away from the spinal cord) nerves, particularly the difference between dorsal and ventral roots leading into the spinal cord (see Figure 4. 1). An important function was recognized for these "roots" in the reflex system (e.g., knee-jerk), which was acknowledged by 1880 to be an exclusive function of the spinal cord. Flourens had studied motor control in a number of species where different parts of the nervous system, including the brain, were selectively destroyed. Using these methods Flourens had shown that the spinal cord is a highly intricate central control sorting out information travelling up and down. In particular, he demonstrated that muscular coordination in ani-

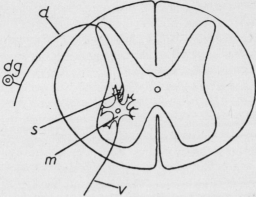

Figure 4-1   (From REFLEX ACTIVITY OF THE SPINAL CORD by Creed, Denny-Brown, Eccles, Liddell, and Sherrington. Reprinted by permission of the authors and the publishers, The Clarendon Press, Oxford.)

mals is localized in the cerebellum. Bernard was interested in the effects of various substances on the physiological function of nerves and is credited, among other things, with showing that curare affects only motor, not sensory nerves.

The English and French schools focused much more of their attention on the structural and functional organization of the nervous system as a whole (as opposed to the Berlin reductionists who were more concerned with the function of isolated nerves). The English and French schools were interested in the processes by which the various components of the nervous system (sensory and motor nerves, peripheral and central nervous systems, spinal cord and brain) were interrelated and integrated. Their experimental procedures tended to be very different from the Germans. The latter often studied nerves outside of the living organism—bathed in special physiological saline solutions, but separated from other living organs. The French and English tended much more to experiment with whole animals, either by careful dissection and/or surgical operation and subsequent testing of motor and sensory responses. While both schools were influential in the subsequent history of neurophysiology, the French and English schools had the greatest direct impact on the physiolgists at the turn of the century: Sherrington, Cannon, and Henderson.

Although much had been learned about nerves and the nervous system in the period prior to 1870 or 1880, there still existed much confusion about how nervous systems function as a whole. For example, it was not at all clear how individual nerve cells connected up to each other in establishing the nervous network. Equally perplexing was how the nervous system sorts out information and channels it in the right direction. Most intriguing but little understood was how the brain and spinal cord coordinate muscular activity, so that antagonistic muscle systems (those that move a part, like the arm, in opposite directions) do not work against one another. Looming above it all, but still baffling to neurophysiologists was, of course, Loeb's question of how the brain determines overall human behavior.

One approach that began to attack the problem of organism-level processes, while still employing the physico-chemical mechanism espoused by Loeb, is exemplified in the study of reflex arcs and their organization. The differentiation of the mechanistic from the holistic materialist's approach can be seen clearly in this area by comparing the work of two pioneers in the study of reflexes: the Russian Ivan Pavlov (1849-1936), and the Englishman Charles Scott Sherrington (1857-1952).

The role of the brain and spinal cord in controlling the output of innate reflex activity, as well as the more complex job of establishing learned reflex behavior, was a question of considerable interest during the mid-and late nineteenth century. The work of Ivan M. Sechenov (1829-1905), in particular, led to the formation of an active and highly influential school of Russian neurophsiology, which reached its highest point in the work of his pupil Ivan Pavlov.

Sechenov was a product of the German reductionist school, having studied between 1856 and 1863 in Berlin with the medical materialists including Johannes Müller, de Bois-Reymond, Ludwig, and Helmholtz. After leaving Berlin, he continued his studies in Paris with Claude Bernard. In the 1860s he made fundamental contributions to knowledge of the cerebral inhibition of spinal reflexes, work to which Sherrington much later paid considerable tribute. Influenced both by the reductionist philosophy and by Bernard's work on elucidating sensory input pathways, Sechenov developed the idea that all behavior resulted from a balance between inflow of stimuli and outflow of response. In a pure reflex action, inflow and outflow were directly and uninterruptedly related as cause and effect. In conscious action, however, outflow was mediated and altered (either intensified or reduced) by the higher brain centers. These centers thus stepped in and modified behavior by interrupting operation of the pure reflex are and affecting the output (motor) signal in a distinct, patterned way. Sechenov's scheme was greatly oversimplified; it represented the kind of model that Sherrington reacted so violently against because of its distorted and specualtive nature. It was through Sechenov, however, that many of the reductionist ideas, especially in neurophysiology, were introduced into Russian physiology. Particularly important in this regard was Sechenov's influence on Pavlov.

Born in Ryazan as the son of a priest, Pavlov studied science and then medicine under Sechenov at the University of St. Petersburgh (now the University of Leningrad). Receiving his M.D. in 1883, he worked for the next two years in Germany, first with the aged Emil Ludwig, and then with Rudolph Heidenhain (1834-1897), who had pioneered studies on salivary secretion and neural connection. Pavlov was appointed to the chair of pharmacology at the Medical Academy in 1890 and to the Professorship of Medicine in 1895, a position that he held until his death.

As a student of Sechenov, Pavlov was influenced toward a highly experimental and mechanistic philosophy. As young student he read enthusiastically the highly popular biological works of Herbert Spencer and George Henry Lewes. While neither Spencer nor Lewes were trained biologists in the usual sense, both were vociferous spokesmen for a materialistic and reductionist philosophy and for a physiological instead of a morphological foundation for biology. Spencer's *Principles of Biology (1864)* and Lewes *The Physical Basis of Mind (1877)* both contain strong arguments for a physiological and molecular approach to all organic phenomena. More important, however, is that both Lewes and Spencer extended their highly mechanistic interpretations to the study of human behavior and to the function of mind. Although Pavlov did not turn to behavioral studies until considerably later in his career (after 1900), he dated his interest in biology in general, and physiology in particular, from an early reading of these works, especially Lewes' *Physical Basis.*

During the first part of his career Pavlov was engrossed in problems of the dynamics of blood circulation and the mechanism surrounding the initiation of secretory activity in digestive glands. In particular, he concentrated on the salivary glands and tried to elucidate the exact nerve pathways by which the glands were innervated. A brilliant operationalist and enthusiastic worker, he was awarded the Nobel Prize in Medicine and Physiology for his work on the physiology of salivary secretion. It was not until 1902, however, that Pavlov turned his research interests away from strict neurophysiology to the study of reflexes and the conditioned (learned) reflex. Pavlov brought to bear on this work not only his experimental skill and mechanistic bias, but also the physiology of reflex phenomena and nervous integration, just coming to fruition in the work of Sherrington.

Pavlov came into the study of conditioned reflexes accidently. He noted that certain routines preparatory to feeding laboatory dogs stimulated the animals to secrete saliva even before food was presented to them. Thus simple, regularly repeated events such as the appearance of the attendant, sounds associated with putting food into feeding pans, or the sight of food all

served as triggers to salivary secretion. Pavlov knew from his own physiological studies that when food is placed directly in an animal's mouth, saliva
begins to be secreted almost immediately. He had attributed this to a reflex
arc whose sensory input was stimulated by the physical presence of food.
The focus of Pavlov's conditioned reflex studies became the question of how
other stimuli only accidentally connected with the presence of food could
serve as a substitute.

In a series of experiments, Pavlov presented a dog with a morsel of food
simultaneously with an unrelated stimulus, such as the ringing of a bell.
After a certain number of repeats of this procedure, the animal would begin
to salivate at the ringing of the bell, even though no food was presented. The
neural pathways involved in any innate (unlearned) reflex response could be
understood in terms of the naturally occurring (genetically determined)
connections that occurred between neurons at each level of the spinal cord
during embryonic development. But the question of importance in understanding conditioning was to find out how new pathways were established
through repetition of events and how they utlimately came to function like
innate reflexes.

Fundamental to Pavlov's thinking about the conditioned reflex was the
concept of temporary neuronal connections made in the cortex by repetition
of external stimuli. Auditory stimuli produced by the ringing of a bell,
repeatedly associated with simultaneous visual stimuli produced by seeing
food, caused a new neural pathway to be formed from the nerve endings in
the tympanum (ear drum) to the output pathways leading to the salivary
galnds. This new connection, a "learned" reflex arc, was made in the cerebral
cortex. Pavlov applied many classical methods of investigating reflex responses to his conditioning experiments. He varied the length and kind of
associated stimulus (i.e., the substituted stimulus, the bell), the length of
time between presentation of the stimulus and the presentation of the
reward, and studied the effect of cerebral ablation on the ability of an
organism to learn new reflex responses.

Pavlov extended his reflex studies to the more complex questions of
learning. He saw the learning process as essentially the buildup of many
conditioned reflex arcs. Memorization, for example, involved the repeated
input of certain stimuli until, by certain associated symbols, a particular
response could be recalled at will. Since learned responses were mediated
always through the cerebrum, Pavlov's learning theory focused on the brain
and its role in the control of conscious behavior. The fertility of Pavlov's idea
and his own indefatigable energy in expounding his views drew to Leningrad
an enthusiastic school of workers who, by the 1920s, opened up a number of
areas of study in the physiology of behavior.

As he grew older, Pavlov's idea on the conditioned reflex became more diffuse and nebulous. The idea of temporary neural connections in the cortex proved considerably easier to postulate than to demonstrate experimentally. Furthermore, many of those concerned with the study of behavior came to recognize that some conditioned responses seemed more stable than others, suggesting that not all of the supposed neuronal connections made in learned behavior were of the same kind. It was also not clear why some responses were acquired after only a few repeats of the stimulus, while others required many repeats. Pavlov's model for explaining complex behavior on the basis of conditioned reflexes came to be regarded by many as an oversimplified picture. Nevertheless, his work opened an extremely important area of study about the relationship between behavior and neurophysiology. To find an experimental tool for investigating this relationship had been the century-old aim of many brain physiologists. In Pavlov's hands at least one component of the learning process could be elucidated in neurophysiological terms.

What Pavlov had shown was that the learned response could be treated in the same terms as inborn reflex arcs, whose anatomical structure (down to the cellular level) and physiological functions were becoming more clearly understood. The significance of this approach was that it provided a specific conceptual framework in which to view a hitherto completely mysterious process. Although the idea of the conditioned reflex was rudimentary, it led many workers to look at learning in a less mystical light. It brought the concept of learning into the realm of anatomical and physiological study. Pavlov's work led to new concepts of psychology, culminating in the input-out put behaviorism of B. F. Skinner.

Pavlov's methods of investigation reflected the reductionist and mechanistic approach that he imbibed as a student. He championed the objective experimental approach to neurophysiological and behavioral problems and strongly denounced speculative ideas. There was no mind-body dualism in Pavlov's thinking, since he saw complex mental processes in terms of simple physical hookups of neuronal pathways. For these reasons, he rejected psychology as a science, claiming that it was based largely on abstract, speculative, and untestable ideas.

Despite his overt mechanism and even reductionism, Pavlov, like Loeb, was also interested in the organism's overall functioning. Pavlov's greatest concern was how the organism's behavioral response could be related to specific stimuli. Although he tried to break down a complex behavioral pattern into its component reflex parts, it was ultimately the animal's behavior in terms of overall response that interested him. Pavlov saw overall behavior of an organism as an integration of many conditioned and innate reflex arcs. The method of integration was unknown; and Pavlov did not have a ready method for investigating it.

Thus the Pavlov school had, from the beginning, a certain built-in-limitation. The mechanistic bias that Pavlov brought to his studies from the German school of physiology meant that any integrative phenomena that could not be understood by simpler analytical methods would be overlooked or misrepresented. The problem of coordination of animal behavior was more complex than a series of simple reflex pathways. Sherrington and others, influenced from the outset against purely mechanistic approaches, sought to avoid the pitfalls of oversimplification. Sherrington's work, to which we now turn, indicates one form of this new direction in twentieth-century physiology.

Charles Scott Sherrington and Nervous Integration

The work of Sherrington was aimed at understanding how nerve impulses are transmitted selectively from one neuron to another in the central nervous system. His work was built on the neuron theory, developed in painstaking detail by the Spanish histologist Santiago Ramòn y Cajal (1852-1934) in the 1880s. Cajal had shown through careful microscopical observation that the nervous systems of all animals—including both the central (spinal cord and brain) and peripheral (nerves leading from the spinal cord to all areas of the body)—are composed of individual neurons, distinct from one another and separated by a gap (later called the synapse). The neuronal theory replaced the once-popular reticular hypothesis, which claimed that the nervous system was composed of a continuous (physically connected) network of fine elements, not strictly cellular in form. The significance of Cajal's neuronal theory was that it provided a sound anatomical understanding of how the nervous system was constructed. On this knowledge an understanding of how messages were passed through the system could be built; it is obvious that control of which impulses went where would be different in a system composed of continuously interconnected elements than in one composed of separate and discrete cells. Using Cajal's findings, Sherrington tried to map the actual pathway a nerve impulse followed from a peripheral receptor (such as a touch receptor on the skin), into the spinal cord and to the brain, and back out over a motor pathway to produce a specific response (e.g., the scratching motion of dogs). The fact that such responses involved the actions of many groups of muscles meant that the incoming impulse had to be fed out over a series of coordinated and interrelated pathways.

Born in England in 1852, Sherrington received his undergraduate degree at Cambridge University in 1884 and, after several years of study in Europe, earned an M.D. from Cambridge in 1893. Introduced to physiological studies as a student by Michael Foster, then at Gonville and Caius College, Sherrington travelled to Germany to study problems of neurophysiology and brain localization. During a two-year stay abroad, he worked in the laboratories of three important physiologists: Rudolf Virchow (1821-1902) and Robert Koch (1843-1910) in Berlin, and Friedrich Leopold Goltz (1834-1902), with whom Loeb had studied at Strasbourg. Although Sherrington was able to see at first hand the kind of work that the mechanistic philosophy produced, he remained skeptical of the apparent oversimplifications to which their models were given, and of the excessive reliance on analogy to purely physical principles. Moreover, he felt that the emphasis on studying isolated parts of total complex systems would never unravel the complex levels of organization involved in brain function or neuronal integration.

Returning to England, Sherrington first took a post as physician, but later moved, first to the University of Liverpool (1895-1912) and then to Oxford as Waynflete Professor (1912-1935). His interest in neurophysiology persisting, Sherrington began his study of nervous control processes by trying to discover microscopically the various neuronal connections in the cerebral cortex. However, he found the material far too complex, so he chose to look instead at a very simple and specific system: a basic reflex arc operating at the level of the spinal cord. Using the monkey as his experimental animal, Sherrington studied the sensory pathways involved in the knee-jerk reflex. His approach, from the beginning, bore the marks of two influences. From his experiences in the laboratories of the German mechanists he drew an aversion to purely reductionist, *in vitro* studies. And from his old teacher, Michael Foster, he learned to regard the nervous system as a whole, functioning unit, the role of the spinal cord being to direct incoming and outgoing impulses over a proper pathways to achieve a coordinated response. Like Loeb, Sherrington set out to study reflex type behavior—automatic and highly regular responses to very specific stimuli. Unlike Loeb, however, he was interested not in a purely physico-chemical explanation for reflex behavior but, instead, in how the organization of the receptors, the central and motor nervous systems, were able to produce coordination.

In his initial studies, Sherrington made two important observations. First, he noted that in any reflex phenomenon there was not only the factor of stimulation, but also that of *inhibition*, a characteristic that had been known since the 1840s. Anatomists had shown that muscles exist in antagonistic pairs—one muscle moving a sturcture forward, the other moving it back-ward. In order for one muscle to contract and produce a significant move-ment, the other muscle must relax simultaneously. Inhibition referred to the means by which one muscle was prevented from contracting when its partner was stimulated. Sherrington pointed out that in the case of simple reflexes, inhibitory control must be exerted at the level of the spinal cord.

Second, Sherrington noted that the best histological evidence available—notably that of Cajal—indicated that neurons did not touch each other physically at the points where their ends came together. Sherrington coined the term *synaps* in 1886 to refer to the space between any two adjacent neurons in a pathway. The "synaps" concept became of fundamental impor-tance to all later understanding of the exact means by which nerve pathways are integrated. If neurons were in physical contact with each other, then it would be possible for a single incoming impulse to be transmitted to a large and random assortment of outgoing pathways with no apparent means of control (an objection brought against the reticular theory). However, if in a synaptic system neuron A has many ramifications going to neuron B, but only a few going to neuron C, then every time an impulse comes over A it will probably stimulate B, but may require repeated stimuli, or simultane-ous stimuli from other neurons, to stimulate C. The pathway from A to B would thus be expected to be used more regularly than the one from A to C. Sherrinton came to the conclusion that even such an apparently mindless system as the knee-jerk reflex showed unified neurological activity. Through his study of this sytem, Sherrington combined behavioral and anatomical knowledge with microscopical studies of the actual connection between the sensor input fibers from the knee and motor output fibers from the spinal cord.

Sherrington pictured the spinal cord in terms of an input-output system (see Figure 4.2), the specific connection made being a function of the anatomical arrangement of neurons leading to, from, and within any particu-lar level. He also noted that reflex arcs at each level of the spinal cord are themselves interconnected through ascending and descending nerve tracts in the outer areas (white matter) of the cord. To make clear his conception of how integration was brought about, Sherrington developed the idea of what he called private and common nerve pathways. Incoming sensory stimuli traveled along a new of private incoming tracts; that is, each stimulus was picked up by one or more receptors in the skin covering the knee cap and transmitted toward the central nervous system. A series of such incoming stimuli were collected by the spinal cord and, through connecting neurons in the central part of the cord (gray matter), were sent out over a common motor

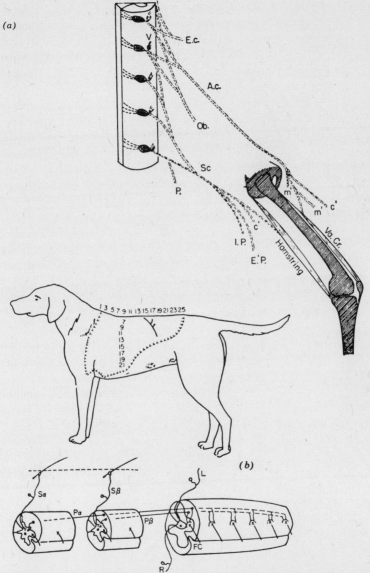

Figure 4-2   (From p. 561 in C. S. Sherrington, "Note of the Knee-Jerk and the Correlation of Action of Antagonistic Muscles, "Proceedings of the Royal Society, 52, 1893. By permission of the Royal Society, London. (Fig. 2b) from Fig. 1 in C. S. Sherrington, "The Correlation of reflexes and the Principle of the Common Path," British Association Reports, 74, 1904; British Association for the Advancement of Science, London. By permission.)

pathway that led to the effector organ (in this case the muscle that caused the leg to jerk when the knee was tapped). The function of the spinal cord, according to this analysis, was to receive a variety of specific impulses and to integrate them into a final common pathway that led to the appropriate effector organ.

Sherrington put together these concepts in his discussion of a more complex reflex system: the scratch reflex in dogs. It was commonly observed that when a dog is scratched at various points on the back or flank, it automatically initiates scratching motion with one hind leg. Because it involved more muscles, the scratch reflex was a better example of the complex neuronal interactions involved in nervous integrations than the simple knee-jerk. Sherrington's basic experimental scheme was simple. He stimulated numerous areas of a dog's back with small electrical stimulators and mapped the region from which the scratch reflex cold be evoked (see Figure 4.3). He then made cuts in various neural pathways leading from these areas of the skin toward the spinal cord and observed how the reflex was modified: was it cut off completely, was it unaffected, or was inhibition removed? For example, he found that cutting one of two afferent pathways from the skin would yield no scratch reflex when that area of the skin was stimulated by subthreshold stimuli. If both afferent tracts were intact, however, what was a subthreshold stimulus for either by itself, would combine to give a threshold response. He explained this occurrence by reference to his concept of the convergence of private pathways into final common pathways.

In 1906 Sherrington went to Yale University to give the Silliman Lectures, which he subsequently published as the epoch-making *Integrative Action of the Nervous System.* In this work he pulled together all his previous ideas on the organism as a functional and integrated whole. He concentrated on the interaction between various reflex levels of the spinal cord, showing how the reflex input at one level could affect and modify the reflex input at another. He pointed out that reflex responses could be diminished or increased by various combinations of stimuli and thus establish a hierarchical pattern of control between one spinal level and another. In other words, he showed that very fine levels of control were possible by interactions occurring within the spinal cord itself and between the spinal cord and peripheral elements.

As the title of the book indicates, Sherrington was particularly concerned with the whole organism and the unity of its response. He maintained that integration could not be understood purely by analytical, *in vitro* studies. In his attempt to emphasize the necessity of studying whole organismic re-

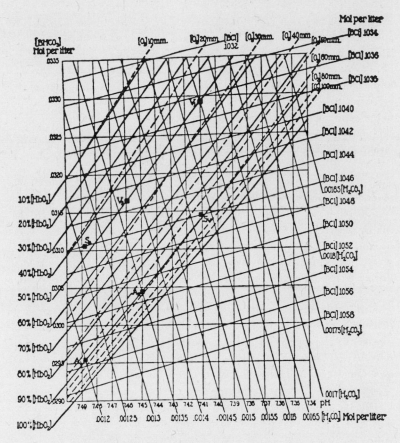

Figure 4-3   *(From BLOOD by Lawrence J. Henderson. Reprinted by permission of Yale University Press.)*

sponses, Sherrington identified three levels at which animal behavior could be approached. The first was the physical-chemical level, involving chemical interactions within individual neurons, producing specific reflex pathways. These processes at the molecular and cellular level welded the body together in what Sherrington called a "unified machine." The second level was the field of the psych, in which many neurophysiological processes were integrated to create a percipient, thinking individual. Simple reflexes did not come to bear at all at this level; they were, however, components of the neurological processes that result in volitional action. The third level of integrative action was the mind-body liaison. To Sherrington, mind was non-physical while body was physical. The third level required that mind and body be compatible elements and that all volitional behavior ultimately reflect this compatibility. In 1906, however, Sherrington saw that the most important direction for neurophysiological research was at the lowest level of organization—analysis of the simple reflex arc. For his work in tracing out motor components of integrated reflex responses and in showing how those components, for example, excitatory and inhibitory responses, were integrated, Sherrington received the 1932 Nobel Prize in Medicine.

Although he felt that his own work was best directed to the simple reflex arc, Sherrington represents that group of physiologists around the turn of the century who shied away from interpreting all complex behavior as merely the sum of numerous individual reflexes. He was, in principle, an antireductionist who reacted strongly against simplified mechanistic views. When he had visted Waldeyer's Laboratory in Berlin, he had been appalled at the great histologist's statement that the brain has a "think-machine" (*Denkmaschine*). He thought that such simple analogies did a great disservice to the complexities of the organism. He was by no means an idealist, who saw mind or "thought" as separate from matter. While he was sympathetic with some aims of the mechanist-reductionist school of the later nineteenth century, he always remained skeptical of its ultimate tendency to reduce all explanations to simple physico-chemical events. Sherrington's own interest in integration forced him to look beyond simple mechanical models. Yet, ironically, he made his most important contributions to knowledge of nervous integration by limiting precisely the levels of organization in the nervous system that he would study and by asking only questions that could be answered experimentally. This was one of the marks of the new mood in physiology: to take from the mechanist-reductionist tradition its rationalistic, experimental side and its conviction that the organism operated in accordance with the laws of physics and chemistry, but to reject the simplistic mechanical models and the insistence that all phenomena had been explained meaningfully when reduced to the molecular level. In this regard, Sherrington represents clearly the other tradition in late nineteenth-century physiology, given so much impetus by the work of Claude Bernard. Purely mechanistic physiology had flourished for a time, especially in Germany, and had been carried into areas of early twentieth-century physiology and biology by advocates such as Loeb. But it was physiology as applied to whole organisms and focussing on the interrelation of parts that came to dominate later twentieth-century physiology.

Internal Equilibrium in Physiological Systems: L. J. Henderson and W. B. Cannon

Sherrington's concept of neuronal integration provided one specific example of how parts of a living system interacted to produce a coordinated effect. But it was not generalizable to a full-scale physiological concept; it remained essentially a neurological development, as seminal as it was in that area. A more comprehensive view of integrated activity, with full physiological implications, came from the work of Lawrence J. Henderson (1878-1942), an American physician, physiological chemist, philosopher, and sociologist. His studies on acide-base equilibria, and later on the buffering action of blood, showed how the many components of a physiological system are interdependent, and how, like Sherrington's reflex arcs, they are all geared to maintain a coordinated and stable life system.

Having become intrigued with physical chemistry in the courses of T. W. Richards at Harvard, Henderson had stumbled on the new electrolytic theory of dissociation of Svante Arrhenius, the same work that had greatly stimulated Loeb. This theory had so excited Henderson that he had written an essay on the subject for an undergraduate contest, emphasizing Arrhenius' own conviction that the theory of electrolytes had direct application to biology. After graduation from Harvard Medical School, Henderson decided to learn more about physical chemistry by studying in Europe, where physical chemistry was becoming an important area of research in the laboratories of the noted colloid chemist, Franz Hofmeister (1850-1922), at Strasbourg. After nearly two years with Hofmeister, Henderson returned to the United States and took up a position as lecturer in biological chemistry at Harvard, where he was to remain for the duration of his long and varied career. Imbued with respect for the German analytical tradition, Henderson was convinced that a biologist could not be concerned only with problems on an organismic or tissue level. Like Loeb, he felt it was necessary to use the methods of physical chemistry to be a biologist and to investigate properly the physical functions of the living organism. Also, like Loeb, the world of physical chemistry—of molecules, atoms, and ions—seemed to Henderson experimentally verifiable, and the ultimate picture of reality. Henderson was, in his own words, at this time a "naive realist," believing that

" . . . The world of science is stable, true and real; that not only the facts and uniformities but also the theories and conceptual schemes are on the whole such that they will endure with nothing more than improvement, refinements, and occasional corrections . . . . "

After returning to the United States in the fall of 1904, Henderson began to study acid-base equilibria, growing out of his earlier interest in Arrhenius' dissociation theory. The relationships that Henderson determined between hydrogen ion (H+) concentration, and the amount of undissociated acid or salt in an aqueous solution led him to describe quantitatively the action of buffer systems. (Buffer solutions through changes in rates of dissociation of particular salts of weak acids and bases, can keep the hydrogen ion concentration (the pH) of a solution constant despite addition of acid or base from the outside.) It had long been known that the body is able to maintain a rather constant acid-base composition (pH), and Henderson concluded that buffer systems must be involved in the blood and tissue fluids. But these fluids were very complex mixtures of many substances and were thus not directly comparable to simple aqueous solutions. Henderson thus turned his attention to simple, artificial, or model buffer systems that contained only a few of the constituents found in blood.

From these early studies between 1904 and 1912, Henderson identified some of the major inorganic buffers that operate in tissue fluids and determined quantitatively their physical and chemical characteristics. Physiological buffer systems, Henderson found, are much more efficient than a simple solution of a weak acid and its salt, the customary buffers used by chemists. He also found that tissue fluids have combinations of various buffer systems: each by itself has a limited buffering capacity, but together they have an enormously increased ability to regulate pH over a wide range.

What excited Henderson most, however, was his observation that the specific buffer systems that the body does use, out of the many it could use, are precisely those that are most suited to pH regulation in context of the body's other systems. For example, the carbonic-acid (H_2CO_3), sodium hydrogen carbonate ($NaHCO_3$) system, only moderately effective *in vitro*, is very effective *in vivo* because the rate of respiration can be varied to increase or decrease the amount of CO_2 in the blood (CO_2 is a product of the dissociation of carbonic acid and specifically triggers the respiratory center in the hypothalmus to increase the breathing rate). Henderson pointed to other similar examples to show how beautifully adapted living systems were to the physical and chemical conditions of the environment in which they had come to exist.

These early studies on buffers brought Henderson to two important realizations. One was that living systems are composed of a complex of interacting factors whose real functions cannot be understood by studying any one component in isolation from the other. The carbonic acid buffer system might appear relatively inefficient as studied *in vitro* by the physical chemist. Only in context of the living system could its real efficiency and physiological role become apparent. Physical chemistry as an analytical tool did not diminish in importance for Henderson, but he came to realize that applied in the purely reductionist manner of analytical chemists and older physiologists, it could lead to oversimplification or erroneous conclusions.

Henderson also realized that a chief characteristic of living systems was the ability to regulate their various processes. As he wote in 1913:

"The right working of physiological processes depends, then, upon accurate adjustment and preservation of physico-chemical conditions within the organism. Such conditions as temperature, molecular concentration and neutrality are now known to be nicely adjusted and maintained; adjusted by processes going on in the body, maintained by exchanges with the environment."

The idea of internal self-controls in physiological processes was not new with Henderson. It had been suggested first in the late writings of Claude Bernard, who had developed the idea that physiological processes functioned so as to maintain a constancy of the organism's internal environment (*milieu intérieur*). A similar emphasis on regulation appeared around the turn of the century in the writings of physiologists such as J. S. Haldane (1860-1936), Sir Joseph Barcroft (1872-1947), and Walter B. Cannon (1871-1945). The recognition of a number of internal control mechanisms led these physiologists to emphasize the study of integration and coordination instead of physico-chemical reductionism as a viable physiological method. It was in the control of processes that living systems could be seen to differ so profoundly from nonliving.

After completing his studies of buffer systems, Henderson applied his methods and principles to the problem of the maintenance of blood pH. The problem was very complex, as he had recognized earlier, but now he felt he had a different approach at hand. What was important to focus on, he reasoned, was not so much the individual chemical reactions (although, of course, more had to be learned about them) but, instead, the *organization* of the components making up the chief blood buffer systems. There appeared, however, to be no satisfactory and quantitative way of doing this, since a change in each component always produced changes in all the others; and how, after all, was it possible to account for a multitude of variables simultaneously? Between 1913 and 1919 (with time out for war-related medical research) Henderson worked on the blood buffer problem. He chose seven interrelated variables to study intensively, gathering experimental

(a)

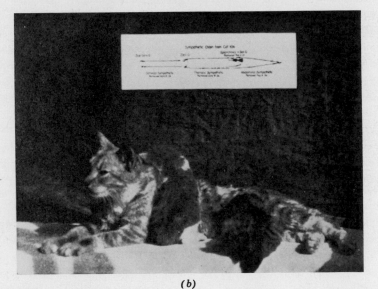

(b)

Figure 4-4   (Illustration reprinted from THE WISDOM OF THE BODY by Walter B. Cannon, M.D., by permission of W. W. Norton & Company, Inc. Copyright 1932 by W. W. Norton & Company, Inc.; Copyright renewed 1960 by Cornelia Cannon.)

data from others as the basis for his mathematical calculations. What Henderson sought was some type of quantitative means of predicting the effects of altering any one variable on the expression of all others. The starting place had to be the physico-chemical data: the effect of oxygen tensions in the blood on carbon dioxide tensions, the effect of these gases on the distribution of chlorides, and so forth. The problem was how to relate these to one another.

With a graphical format known as a Cartesian nomogram (see Figure 4.4), which he discovered almost by accident, Henderson finally found a means of accounting for the data represented by his seven variables. The nomogram shows the effect of changes in any one variable on all of the others, thus providing a means of representing complex interactions of multicomponent systems.

Henderson presented his detailed researches in a series of 10 papers titled, "Blood as a physico-chemical system," published between 1921 and 1931. In these papers, and in the Silliman Lectures on this subject at Yale in 1927 (later published as *Blood: A Study in General Physiology*), he showed not only how his mathematical methods could be useful for analyzing complete systems such as blood buffers, but he also emphasized the importance of applying this approach to other physiological systems.

Henderson's methodology started from the basic assumption that all living systems function according to the laws of physics and chemistry. He thus derived part of his inspiration from the same physico-chemical materialism that had so stimulated Loeb. Buffer systems were not mysterious, nonphysical entities, but specific chemical reactions following the dissociation laws of physical chemistry. Henderson was definitely a materialist. But studying individual reactions was not enough; the blood buffer system could not be understood by simply reducing it to a list of separate parts. What was crucial in the new method was the focus on *organization* and on finding a quantitative means of describing it. He thus derived a second part of his inspiration from the physiological schools of France and England, which emphasized the interaction of parts of the whole and were less committed to the physico-chemical reductionism of the medical materialists. Henderson paid considerable tribute to the experimental, materialist, and mechanistic school: he frequently cited the work of Pavlov, Roux, Driesch, and *Entwicklungsmechanik* for their rigorous approach to biological problems, and even for their emphasis on the organism's ability to regulate its own processes. He also paid tribute to Sherrington and especially to Cannon for their concern with integrated processes. Most clearly, however, Henderson's emphasis on regulation brought him to a full recognition of the importance of Claude Bernard's concept of the *milieu intérieur*, introduced in 1878 in the posthumously published work *Lecons sur les phénomènes de la vie communs aux animaux et aux végétaux* some 50 years earlier. More, perhaps, than any other person, Bernard's influence can be seen in the twentieth century through the work of Henderson. Henderson became a great champion of Bernard's work in the United States, and was responsible for the first English translation of the *Introduction to the Study of Experimental Medicine* in 1927. The important principle behind the work of both Henderson and Bernard—the body's means of maintaining a constant internal environment—was given a precise designation in the 1920s by Henderson's colleague at the Harvard Medical School, Walter Bradford Cannon. Cannon extended Henderson's quantitative regulatory concepts to cover not just the tissue fluid system, but the whole body. Although his work was more empirical and less mathematical than Henderson's, it was no less concerned with the interaction of many parts in a complete and integrated physiological system.

During World War I, Cannon, a physiologist and physician trained, like Henderson, at Harvard, was employed by the U. S. Army Medical Corps to work on the problem of shock—the physiological condition in which the body is unable to continue maintaining itself, the various systems collapsing because of their inability to check runaway processes. Cannon saw clearly that shock resulted from the breakdown of regulatory mechanisms—the rapidity and extent of the breakdown once it starts indicating to him the importance that these control systems play under normal conditions. He recognized that regulation of overall bodily processes such as temperature, metabolic rate, blood sugar level, heart and breathing rate was achieved not simply by on-the-spot substances such as Henderson's blood buffers, but also by the interaction of nervous and endocrine systems. Cannon proceeded in his own work to study initially one component of this total complex, the sympathetic branch of the autonomic nervous system. The autonomic nervous system is the part of the nervous system of vertebrates which regulates involuntary responses—especially those concerned with nutritive, vascular, and reproductive activities. It is composed of two branches, the sympathetic and the parasympathetic systems.

Cannon discovered that the sympathetic system is the major "foreman" that virtually controls all the other regulatory systems in the body. For example, when the body is heated up (because of exercise or external temperature) the sympathetic system triggers several responses:

1. Muscles along capillary walls near the skin surface relax so that more blood is allowed to flow near the surface and thus release heat.
2. The sweat glands are stimulated to release water that evaporates and thus carries away heat with it. In a similar manner, when body temperature falls:
a. Capillary muscles constrict, thus less blood flows to the surface and less heat is lost.

b. Epinephrine (adrenaline) is released from the adrenal glands (under sympathetic control) into the blood, accelerating the body's metabolic processes, which produces heat. All of these interactions, Cannon pointed out, serve the primary function of maintaining the temperature (of warm-blooded animals) as nearly as possible at a constant level.

In a series of remarkable experiments, Cannon removed parts of the sympathetic nervous system from animals so that certain regulatory mechanisms were no longer operative (e.g., sweating). He found that even animals in which the entire sympathetic system was rendered inoperative were perfectly able to continue living and functioning (females could even conceive and give birth) under normal laboratory conditions (see Figure 4.4). However, if the experimental animals were put under any stress (e.g., lowered or raised temperature, physical exertion, or lack of food), they succumbed much more readily than a normal animal. In other words, they could not adjust their internal system to change and thus could not maintain the body fluids at the delicate balances of sugar, temperature, and ionic concentration required for life. Animals without the sympathetic nervous system, which controls these processes, must depend completely on the external environment for all forms of regulation. The internal environment has no means of buffeting itself against changes from the outside.

Cannon's neurophysiological studies on the sympathetic system, along with his related studies of endocrine function, led him to coin the term "homeostasis" to describe the self-regulating processes by which the constancy of the internal environment is maintained. The term consisted of two Greek roots: "homeo," from "homio," meaning "like," and "stasis," meaning "fixed condition." Thus Cannon used the term "homeostasis" to describe the internal environment of the organism as "similar to a fixed condition." He was careful to point out that homeostasis means "like," not "identical to." One of Cannon's most important contributions in defining the concept of homeostasis was to point out that the constancy of the internal environment is maintained, not by sealing the organism off from its environment, but by regulating ongoing physiological processes. The equilibrium thus achieved is dynamic: the component parts are constantly changing (ions are coming in and going out, heat is constantly being generated and lost), but the overall system stays the same. As Cannon pointed out, dynamic equilibrium is maintained by regulating the *rates* of physiological processes, slowing them down or speeding them up as changing conditions demand. Such an arrangement was able to provide a much more sensitive and delicate means of regulation than one that involved simply turning on or off a whole process.

Although Cannon saw control mechanisms serving to keep a constant fluid environment for the body's cells and, although each control mechanism functioned in accordance with physio-chemical laws, the organization of control mechanisms was supracellular or supramolecular. A whole system of nerves, endocrine glands, various muscles, and the like were integrated, each part affecting the functioning of others. Like Henderson, Cannon did not believe that a reduction of control processes to single molecular interactions would depict adequately how sensitive regulation was maintained. The organization and interaction of parts was what differentiated living systems from simple physical or chemical processes. He thus saw the organism as a whole—each part having its own functions, but nevertheless integrated with all other parts by means of the various control processes that maintained the internal fluid environment.

Cannon's and Henderson's work, taken together (along, perhaps with that of J. S. Haldane on the regulatory mechanism of breathing rate) represents the most full and influential statement of regulatory, equilibrium concepts in early twentieth-century biology. Both Cannon and Henderson focused on the chemical concept of dynamic equilibrium, but they went beyond the simple chemical systems to show how *organization* of various equilibrium systems brings about a precise and highly efficient regulation of the organism's many interacting systems. Both considered themselves disciples of Claude Bernard and, through their Harvard association, exchanged ideas easily and continually. Cannon's work in particular profitted much from developing techniques in general—especially in neuro-physiology—during the 1920s and 1930s. This included *in vitro* studies of growing nerve cells, development of the cathode ray oscilloscope and its use in studying nerve conduction, chemical analysis of synaptic transmission, and growing awareness of the specific chemical role of hormones. And Henderson's work in particular would not have been possible without the development of concepts in physical chemistry and techniques such as blood gas analysis developed in the late 1890s and early 1900s.

Probably the most influential aspect of Cannon's and particularly Henderson's work was the new direction they gave to physiological work and, by analogy, to other areas of biology. This method involved a departure from the strict mechanist-reductionist approach that had remained a strong part of physiology from the 1850s onward. By asserting their belief that all parts of any living system operate on the basic laws of physics and chemistry and by emphasizing the fundamental physiological importance of organization, they avoided the pitfalls of vitalism and metaphysics on the one hand and of reductionism (with its denial of the importance of organization as anything separate from interacting molecules) and naive mechanistic materialism on the other. Their new methods (the nomogram for Henderson) and concepts (homeostasis for Cannon) provided a way in which the whole organism could be studied by the experimental, quantitative, and mathematical methods that were considered components of rationalistic science. Henderson and Cannon were holistic materialists who argued from a position of belief in material causes, but yet sought the interconnections between component parts of a system.

The work of Loeb, Sherrington, Henderson, and Cannon was distinctly antivitalistic and was often criticized by the bastions of vitalistic thought that still held forth in the 1920s and 1930s. Loeb's conclusions had been so brash and so extreme, however, that vitalists could respond in a way that others, not necessarily vitalists themselves, could still follow and agree with. But the latter three men were not in that position. Vitalists found it more difficult to introduce a vital force into complex regulatory processes when the organization of those processes was shown to be achieved through the interaction of many physico-chemical systems. But they did argue, nonetheless; men such as Hans Driesch, Henri Bergson, J. C. Smuts, and Ludwig von Bertalanffy. Although they disguised their vital force in other names, the real meaning of their arguments became more clear as the years wore on. Vitalism as a doctrine was on the run, and physiology, along with embryology and heredity, was the leading the pursuit. There was, in fact, another alternative, which we have called holistic materialism, that was both rational and in the long run made most sense of the data at hand. All biological processes are, in fact, complex interacting systems, and to study life meant to study the organization of these interactions. The era of naive mechanism or metaphysical vitalism was past. Biological thinking had begun to move toward a more sophisticted stance.

The Distinction Between Mechanistic and Holistic Materialism

The integrative approach to physiological problems began to take precedence over the mechanistic approach by the 1920s. This change was more than the replacement of a simpler theory by a more complex one. It was a replacement of one philosophy of biological phenomena by another: the replacement of mechanistic materialism by holistic materialism. Although not called by that name, it amounted to a new philosophical view given explicit formulation as "organic mechanism" by the philosopher Alfred North Whitehead (1861-1947). Whitehead's views on this subject were published in his 1925 Lowell Lectures at Harvard (as *A Philosophical Interpretation of Nature*), which Henderson reviewed the following year in *The Quarterly Review of Biology*. Implicit in Whitehead's analysis was the attempt to bridge the gap between mechanistic materialism and all-out idealism (in whatever form: vitalism, "organicism," "emergence," and so forth). The means of doing this was to emphasize not material units in science so much as in "events." The event was "the ultimate unit of natural occurrence." Every event is potentially related to all other events, and thus no event is independent itself. In Whitehead's scheme, living organisms should be studied not so much as combinations of material parts (i.e., not as structures) so much as combinations of processes. Although Whitehead's position may have appeared nonmaterialistic in its suggestion that parts have different characteristics separately from those they possess when combined into a whole, it was not a return to vitalism. It was one expression of the same holistic materialism that had replaced the mechanistic materialism of the previous century. Whitehead's view greatly stimulated Henderson. It fit well into his own research problem, physical chemistry, where atoms and molecules as material entities are less important than the *interactions,* the processes, in which they are involved.

In viewing the differences between the mechanistic and the holistic materialists, confusion frequently arises. The chief, or important, difference was not that, by and large, the two groups worked at different levels of organization within living systems. It is true that many, if not most, mechanistic materialist were concerned with the cellular or physico-chemicals, electrical currents, ion changes, and the like. It is also true that chemical, electrical currents, ion changes, and the like. It is also true that most holistic materialists worked at higher levels of organization: tissue, organs, or systems. By definition, the latter group was often concerned with more complex interactions, including more component parts. But this difference in level of focus is not the main distinguishing feature of mechanistic from holistic materialists. After all, workers such as Henderson focused on the physico-chemical level, although not in a mechanistic way; and Pavlov focused on the organismic level in a mechanistic way. The difference is one of outlook, not primarily of level.

Very few of the classical mechanistic biologists, including Loeb himself, ever maintained that studying physico-chemical interactions, to the exclusion of organ or tissue level interactions, would give a complete description of the animal (or plant) organism. What they did maintain was (1) that the level of chemical interactions was the most fundamental, determining all the higher levels of organization, and (2) that from knowledge of the physico-chemical interactions, all higher level interactions could be predicted and, eventually, experimentally confirmed. There was little room in the mechanists' philosophy for higher-level interactions that could not be deduced in principle directly from lower-level interactions in a summary way. It was a philosophical position that maintained that the whole was equal to the sum of the parts. Knowing all the parts, each studied in isolation (i.e., *in vitro*), one had to put the parts together in order to understand the whole. No mechanist would have denied that some attention and even some techniques might have to be developed in order to put the parts together. The general methods of synthesis and analysis are never identical or equivalent. But it was not primarily in the use of analysis or synthesis as the main mode of operation in which the old mechanists can be distinguished from the new holistic physiologists.

Although there are some similarities, there is a real philosophical distinction to be made between the work of a Loeb, on the one hand, and a Sherrington, Cannon or Henderson, on the other. All were materialists; they believed that biological phenomena were the result of the interaction of atoms and molecules in accordance with discoverable (although not necessarily discovered) physical laws. The mechanists differed from the holistic materialists in how they viewed studying interactions. To mechanists the properties of the whole were derivable from the properties of the separate, individual parts. To holistic materialists, the properties of the whole were derivable partly from the properties of the separate parts, but also from their properties when acting in consort. According to holistic materialists, there is a set of properties belonging to any component of a whole that can be described only when the component is interacting with other components; those are the properties that are conferred on the components by the act of interaction itself. For example, the integrative control of antagonistic muscle parts could only be described in terms of pathways of stimulation *and* inhibition. To study either stimulation or inhibition pathways alone would miss the integrative (*i.e.*, interconnected) properties characteristic of both pathways. Similarly, the control properties characteristic of the autonomic nervous system could not be deduced by separate study of just the sympathetic or parasympathetic systems. And finally, the total buffering potential of the blood could not be understood as the additive effects of each component buffer system described separately. A complete description necessarily required that the study of interactions be included along with the study of separate parts.

The difference between mechanists and holists is made precisely clear on the latter point. Mechanists such as Loeb would have said that, of course, one must study the interactions that occur at higher levels (than the molecule); he would not have claimed that by studying only nervous inhibition one would automatically be able to predict all the characteristics of the integrated nervous system, or that, having described the sympathetic system, one would automatically be able to predict the characteristics of the parasympathetic system. But he would have said that one could theoretically describe all the properties of the sympathetic system necessary to predict how it would interact with the parasympathetic system by studying each separately. The point of difference is just this: to a mechanist, interaction does not impart new characteristics to any one component when it is interacting as part of a whole or when it is acting in isolation. A complete description of the characteristics of any part is possible by studying that part separately from others. Mechanists believe that studying interactions is also important. But the interactions are just quantitatively more complex situations. To holistic materialists, on the other hand, new characteristics emerge from the interaction of parts. These new characteristics are not merely quantitatively more complex. They are qualitatively different. New characteristics of parts emerge during interactions. These new characteristics result from the parts affecting and even altering each other. To holistic materialists, these new characteristics are interpretable in terms of rational laws of science, but they are not quantitatively extrapolatable from analytical studies alone.

Although holistic materialism provides, ultimately, a more complete and accurate description of reality, mechanistic materialism has had a very important historical role to play in the development of the various sciences, especially biology. The mechanistic approach has, in virtually all the sciences, been the first in time to replace idealistic thinking; hence it has generally represented a more accurate way of relating theory to the real, material world. And, too, the mechanistic approach has often been the only practical way to begin the study of a complex process. However, the limitations of mechanism become apparent when it becomes the sole guiding principle of investigation; certain types of interactions, for example, nervous integration, cannot be understood accurately by a process that looks for the functioning of the whole solely in the properties of its separable parts.

Biology and Society: Origins and Metamorphosis of Mechanistic Philosophy

What accounts for the widespread development of materialism in general, and mechanistic materialism in particular, throughout biological science from the 1850s (in physiology) through the 1880s and early 1900s (in other areas of biology)? An attempt to answer this question leads to an examination of the general cultural—philosphical, economic, and political—conditions of this period. Application of the mechanistic philosophy throughout biology and the ultimate replacement of mechanism with holistic materialism was related to, and influenced by, developments in a number of areas of human affairs: the physical sciences, social sciences, and the arts. Science is not divorced from the cultural period in which it develops. Any attempt to study science in total isolation from its historical period gives credence to the all too prevalent myth that science is "neutral," and that scientists exist in the "ivory tower." The history of science shows continually how scientists are products of their times, directly and indirectly; their very philosophical positions reflect the material conditions of the social order in which they were raised. In a thought-provoking book, *A Generation of Materialism,* Carlton J. P. Hays has demonstrated the prevalence of a materialistic philosophy, more precisely a mechanistic materialist philosophy, in the two generations before World War I. As Hays points out, it was during this period that European political leaders (e.g., Bismarck and Cavour) redefined national boundaries for central Europe on strictly economic and geographic (materialistic), instead of on cultural or ethnic (idealistic) lines; that Marx made a political philosophy out of materialism, and that Herbert Spencer raised the idea of Darwinian struggle to the status of a social ethic. A materialistic ethic was conducive to the development of a strictly mechanistic and reductivist biology. World War I became a dividing line in politics, philosophy, and science in the early twentieth century. It marked clearly and unmistakably the triumph of materialism and the death of idealistic philosophy as a source of ideas for governing human society. The old-world order of Europe, which had fostered divine rights, feudalism, cultural nationalism, and an agrarian economy, gave way to principles based on pragmatism, capitalism, *Realpolitik*, trade unionism, and socialism. Life, and the principles that guided it, were qualitatively different by 1920 than they had been in 1900. Change in fundamental approaches to problems was in order, a movement in which the sciences, particularly biology, took part.

What was behind this *milieu*, how did it become established, and what were its causes? To establish parallels in any period as, for example, between biology and physics or between science and philosophy is not necessarily to establish causal connections. If the development of mechanism in biology was fostered by the same conditions that fostered materialism in philosophy and politics, then the important historical problem becomes to seek out those conditions.

The post-Darwinian period saw the enormous expansion and increase of industrializatión throughout Europe and North America, but principally in England, France, Germany, Italy, and the United States. Increased industrialization had enormous economic consequences for these countries; it created (1) the need for continued sources of raw materials, (2) the necessity of insuring markets for the products, and (3) a growing proletariat class within each country that was more and more the victim of the machine age and its economic and psychological consequences. These factors all played an important part in fostering the mechanistic conception articulated so well by intellectuals in the late nineteenth and early twentieth centuries.

The rapid acceptance of social Darwinist ethics in the nineteenth century coincided with economic trends associated with industrialization and capitalism. As an economic system, capitalism fostered the ethics of competition and exploitation (either of resources or people, or both). Growing mechanization and industrialization intensified the conditions under which these ethics came into play. If mechanization and the exploitation of labor were the rule in one industry, others—the competitors—were forced to follow suit or fail in the competition. As large industries developed during the nineteenth century, exploitation of labor became the chief means of increasing profits (child and woman labor, etc.). When social outrage was sometimes raised against the industrialists, they responded with Spencer's ready-made Darwinian ideology. Since competition was the basis of all biological organization, it was also regarded as the basis of human social organization. The results of competition—actual survival and the quality of life—were dished out according to fitness. Darwinian populations in nature fought it out—they were, in Tennyson's words, "red in tooth and claw." The organisms that survived were, by definition, the most fit. In a similar vein, social Darwinists argued that if in human society certain groups, for example, the industrialists, were at the top, they deserved the privileges that they enjoyed because they were, by virtue of their material successes, the "most fit." On the other hand, workers were at the bottom of the social and economic ladder because they were least fit. A life of degradation, disease, and ignorance was considered to be the natural state in which the poor should live; it was, after all, only a tangible indication of their unfitness. In the United States the new industrial magnates—the Carnegies, Rockefellers, and Vanderbilts—who were high on the wave of their own economic success, justified their exploitation of masses of people by the Social Darwinist ethic.

Like Darwin's original theory, Social Darwinism was a mechanistic-materialist philosophy. Life was combat, in which the victor obtained material rewards and the loser suffered material losses. On two accounts, Social Darwinism was fostered by the economic and political-social *milieu* of the later nineteenth century (1) as a justification for the exploitative tactics of capitalist industry, and (2) as a highly mechanistic (or materialistic) philosophy of social process, focusing on material gains and losses and seeing human life as a constant and natural struggle for survival.

Psychologically, a new conception of human worth and the purposes of human life grew up simultaneously with the industrial boom of the latter half of the nineteenth century. The growth of factories and the widespread mechanization of industry brought large groups of people into the cities and into contact with a highly (for the period) mechanized way of life. In the factories men worked in close contact with machines, or indeed they worked in a highly mechanized process in which they functioned as if they were machines. Labor was measured in mechanical terms—in man-hours—or as indexes of productivity and not in terms of craftsmanship or quality. In the cities working people often lived as animals, in subhuman conditions where life was cheap. It is no accident, in any period or in any country where mechanization has made major inroads, that idealistic philosophy has been replaced in the most basic sense by a materialist philosophy. People were replaced by machines and treated as adjuncts to machines. Under these conditions a philosophy of mechanistic materialism served to rationalize the machine as a model for natural order.

In a similar manner, the metamorphosis of mechanistic into holistic materialism observed in biology from the 1920s onward was paralleled by a similar metamorphosis in society at large. Bismarckian politics of balance (static) relationships among big powers began to give way to the philosophy of dynamic equilibrium, of shifting, constantly changing spheres of political and economic influence. It was recognized that treaties, the bulwark of Bismarck's political system, could not reflect changing social, economic, and political conditions. In place of static and old-fashioned "pay as you go" capitalism, Keynesian economics introduced the concept of deficit spending based on the notion of internal regulation through processes that restored equilibrium. Keynesian economics was based on the idea of a constant internal environment; on the idea that economic systems had to maintain equilibrium against constantly changing external conditions. The older *laissez-faire* form of capitalism had specifically avoided such regulation—there had been no notion in classical economics of constantly changing equilibria. The sole regulatory process had been, by and large, the law of supply and demand.

A physiologist such as Henderson was philosophically convinced of the application of equilibrium concepts to economic processes and held the Keynesian system as a superb example of nonmechanistic but still materialistic economics. Interested in sociology, Henderson taught a course on contemporary sociology at Harvard in which he made clear the connections that he saw between chemical and economic equilibrium. The equilibrium that both Keynes and Henderson promulgated was very different from that championed by Helmholtz or Bismarck. The latter was static, the former dynamic and constantly changing. Henderson was greatly influenced, in a direct way, by the Italian sociologist and economist Vilfredo Pareto (1848-1923) who, in 1916, had applied the concept of dynamic equilibrium to social systems. Pareto and Henderson believed that societies, like the organism, operate through many interacting systems, all of which can regulate themselves or each other. Self-maintenance was a characteristic of all systems, from blood buffers to the central nervous system to large-scale societies. To Henderson, Keynesian economics was only a conscious manifestation of the regulatory principle that operated within all dynamic systems. But what Henderson most liked about the Keynesian system, as compared to simple supply-and-demand economics of the classical school, was that it recognized the constant interaction of numerous factors. Classical supply-and-demand theory was highly mechanistic; it ignored numerous economic situations and saw all processes as controlled by a single mechanism. It ignored the fact that supply is changed by demand, often in ways that produce spiral, runaway phenomena (recessions, crises, or depressions). Through his sociology course at Harvard (in the 1930s), Henderson imparted his views of equilibrium and holistic materialism to a whole generation of American sociologists, including Talcott Parsons, George Homans, and Crane Brinton.

While Keynesian economics represented a departure from the classical, highly mechanistic materialism associated with early capitalism, it was not a complete break with the mechanistic tradition. It failed, for example, to see that debt and deficit spending have some upward limitation in the real world. It was a departure from simplistic mechanism, substituting a more complex mechanistic conception that took into account more interactions among the components of society.

More directly related to changing economic patterns and the shift away from mechanistic materialism in science is the enormous growth during the twentieth century of communist societies throughout the world. Marxism as an economic system is based on the philosophy of dialectical materialism (another name for the general philosophy of holistic materialism), itself a philosophical system born out of direct opposition to mechanism in the early and middle parts of the nineteenth century. The philosophical mood was changing throughout the world in the 1920s and 1930s; mechanism was on the decline. Philosophers, political and economic leaders, and scientists were recognizing that no understanding of the real world could be complete without developing methods for studying the interactions of components in any system. No one would claim that a shift toward this view in physiology was caused directly by a parallel shift in economics or philosophy. Nonetheless, such parallel shifts are not totally unrelated. The relationships can be seen as coming about in two ways. (1) If events in the economic as well as physiological world are really interrelated, then attempts to understand the reality in each case will lead, independently, to a gradual recognition of those relationships. Thus approaches to reality in two different areas of thought will develop along similar lines because both will reflect the same reality in the material world. (2) General cultural views influence the kinds of questions that people ask, even in specific areas such as biology that appear to have little in common with philosophy, economics, or other "external" influences. A person exposed to machine analogies in his or her education is likely to think in terms of such analogies in other areas; it is not unusual that prevailing modes of thought in any period tend to have a certain compatibility with one another. Thus Loeb and other more simplistic mechanists were products of an age in which *laissez-faire* economics was still dominant; it is not unusual that Loeb's mechanism took a *laissez-faire* form. It lacked a sense of complex interactions in which the parts themselves are changed as a result of the process itself; it lacked the idea that the process itself changes, evolves, over time; it was as static a view of organisms as *laissez-faire* economics was of human society. The newer views, dialectical materialism in philosophy, Marxism in economics and history, and holistic materialism in biology, emerged partly because mechanism could not account adequately for complexities encountered in the real world.

In history, parallels do not automatically indicate cause and effect relationships. To discuss the origin of the mechanistic view and its replacement by holistic views requires considerably more study than has been carried out at the moment in cultural history. Discussion of the above parallels is meant less to suggest the specific lines of cause and effect than to suggest that such parallels are not fortuitous. Science and society have many points of interaction, or interconnectedness. It is toward a deeper understanding of the interconnections, particularly in terms of directions of influence, that the history of science can and should in future years direct more of its attention.

CHAPTER V

The Convergence of Disciplines: Embryology, Genetics, and Evolution, 1915-1960

THE GIANTS OF post-Darwinian biology, Ernst Haeckel and August Weismann, had dreamed of a unified approach to the study of life in which all biological phenomena would be seen as intricately related. Embryonic development was to be explained in terms of heredity, cytology, biochemistry, and even phylogenetic history. Phylogeny, in turn, was to be understood only in terms of heredity, embryology, and cell physiology. Their dream was premature; the only unified approach that Haeckel and Weismann could offer was based almost wholly on speculation. By the 1920s, however, that dream was more of a reality. The revolt from morphology was complete, and the introduction of experimental methods into embryology, heredity, and cytology, along with the rise of the mechanistic philosophy, had suggested that all biological phenomena could be understood in terms of chromosomes and genes on the one hand, or molecules and atoms on the other. Many biologists began to see that not only in content but also in methodology the diverse branches of biology had more in common than had previously been thought.

Central to the unification of areas apparently as diverse as cytology, cell physiology, development, and evolution was the study of genetics. Knowledge of chromosome movements during formation of egg and sperm or during embryonic cell division was the contribution of embryology and cytology. Knowledge of the structures of chromosomes and their relation to the factors of Mendelian heredity was the contribution of genetics. Knowledge of the nature of individual adaptations and their changes in frequency within a population over time was the contribution of evolutionary theory.

The convergence of disciplines witnessed in the 1920s and 1930s occurred primarily within two large areas. On the one hand, embryology, biochemistry, cytology, and genetics began to come together to form a unified, cellularly and physiologically oriented view of development. On the other, biometrics, Darwinian evolution, field natural history, and classical Mendelian genetics converged to provide a rigorous, quantitative and, for the first time, logically consistent theory of the origin of species. These syntheses were something more than the summation, the addititive or cumulative effect of ideas from different disciplines. In coming together, ideas from one area transformed those from another, and the unified approach that emerged was qualitatively more than the sum of its parts. This was especially true in the area of evolutionary studies—population genetics was something qualitatively different than either Mendelian genetics or the population concept of field natural history that comprised it. But it was also true of embryology: Spemann's supracellular, or Paul Weiss' field theory, approach to embryonic development were qualitatively different from older theories dealing simply with greater numbers of cells or from Roux's mechanistic mosaic theory of differentiation. The fragmentation of the older, more unified biology of Haeckel and Weismann had occurred in the first decades of the century in conjunction with the rise of mechanistic philosophy. The reunification began to occur after 1920 with the rise of the holistic philosophy as espoused by workers such as Sherrington or Cannon. In the convergence of this later period, genetics, evolution, and embryonic development began to remerge in a new, and less speculative, unified theory of the living system.

This chapter first traces the unification of genetics, cytology, and embryology and then that of field natural history, biometry, and genetics. The further unification of these two large subgroups did not come until somewhat later, with the rise of molecular biology in the 1950s. It will consequently be dealt with in the following chapter.

PART I: The Embryological Synthesis

Hans Spemann and Entwicklungsmechanik

A number of unresolved questions grew out of the work of Roux, Driesch, and the *Entwicklungsmechanik* school—and many if not most of these remained unanswered at the turn of the century. Most of these questions centered around the problem of embryonic differentiation, but they took a number of forms. (1) Was differentiation caused by factors internal to the embryo, or was it regulated by external factors (what Roux called independent or dependent differentiation, respectively)? (2) If differentiation was caused by external factors, what were they? (3) If it was caused by internal factors, what was their nature? Was it, for example, a result of the qualitative dividing up of genetic material into successively more limited types as embryonic cells divided? Or was it caused by some internal mechanism that caused some cells to express some traits, out of a full complement, while others expressed other traits? Deriving from these questions is one of great significance to embryologists: what guided the stepwise and precise sequence of events characterizing embryonic growth from fertilized egg to adult?

To answer some of these questions embryologists in the early twentieth century turned toward the study of the events leading to the formation of various embryonic tissues and organ systems. Lacking the precise tools for studying these problems on a molecular and biochemical level, embryologists focused on levels of organization more readily accessible to the observational techniques of the day. Among the leaders of this approach to embryology was the German, Han Spemann (1869-1941). A strong proponent of the *Entwicklungsmechanik* tradition, Spemann attempted to approach the later development of the embryo by the same rigorous and experimental techniques that Roux, Driesch, and others had applied to earlier development. Spemann's concept of embryonic induction became one of the most important theories in twentieth-century biology, and his later theory of "the organizer" one of the most controversial.

A director of the Kaiser-Wilhelm Institute from 1914 to 1919, Spemann became Professor of Zoology at Freiburg in 1919, a post he held for the remainder of his active career. Before the turn of the century, Spemann had begun investigating the developmental processes leading to the formation of the lens in the frog eye. The lens itself develops from ectodermal tissue (the outermost layer of the embryo), which lies just outside of and next to the optic cup, itself an outgrowth of brain tissue (see Figure 5.1). In a beautiful set of experiments in the ealy 1900s, Spemann showed that a direct and dependent relationship existed between the presence of the optic cup and the formation of lens tissue from the overlying ectoderm. If presumptive optic cup tissue were removed before it came into contact with overlying ectoderm, the latter did not develop into a lens. Conversely, if differentiated optic cup tissue was transplanted to a region of the embryo where it normally did not develop (i.e., just under the ectoderm of presumptive tail region) it caused the overlying ectoderm (which normally becomes skin tissue) to differentiate into a lens. From this work, Spemann concluded that the optic cup *causes* the differentiation of lens tissue from any ectodermal tissue in contact with it. In some way an influence must spread from the optic cup itself into the neighboring tissue to trigger its differentiation. This causal process Spemann called *induction*.

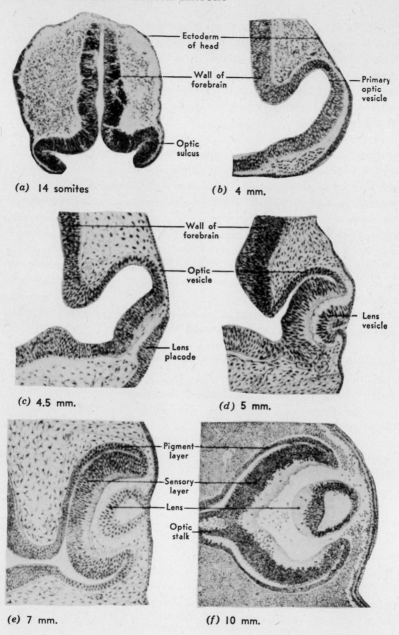

(a) 14 somites

(b) 4 mm.

Ectoderm of head

Wall of forebrain

Optic sulcus

Primary optic vesicle

(c) 4.5 mm.

(d) 5 mm.

Wall of forebrain

Optic vesicle

Lens placode

Lens vesicle

(e) 7 mm.

(f) 10 mm.

Pigment layer

Sensory layer

Lens

Optic stalk

Figure 5-1 (From THE FOUNDATIONS OF EMBRYOLOGY by Bradley M. Patten. Copyright 1958 by McGraw-Hill Book Company. Used by permission of McGraw-Hill Book Company.)

Spemann saw that there was not just one but a whole series of stepwise inductive processes at work in the development of any complex organ. In eye formation, for example, the head mesoderm induces formation of the brain, the midregion of the brain induces the formation of the optic nerve and vesicle, and the optic vesicle induces the formation of the lens. In each sequence the induced tissue from one step becomes the inducer for the next. The induction process thus established a mechanism for the orderly sequence of events in differentiation.

If the inductive process were organized as he postulated, Spemann reasoned that somewhere in the developmental process there must be a primary inducer that initiates all the subsequent responses. With his graduate student, Hilde Mangold, Spemann performed a series of remarkable experiments in the early 1920s that provided some clue about the existence of a so-called "primary inducers." Spemann and Mangold trans-

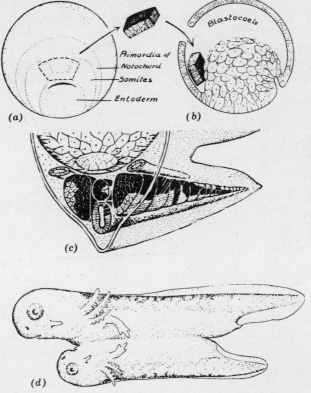

Figure 5-2 (From Willier, B., Weiss, O., Hamburger, V.: ANALYSIS OF DEVELOPMENT. Published by W. B. Saunders Company, Philadelphia, 1955. Used with permission.)

planted bits of tissue from one region of a young (blastula) embryo to a different region on another young (blastula) embryo. Like Spemann's earlier transplantation of optic cup tissue to the tail region, the transplant would induce changes of only a local sort; for example, a single specific organ would be induced in the area of transplantation. However, when Spemann and Mangold transplanted a region from a part of the embryo known as the dorsal lip of the blastopore (a part of the gastrula, as shown in Figure 5.2), some very remarkable results occurred.

The period of the 1920s was one of great excitement and hope in embryology, in which a number of specific questions, deriving primarily from the organizer concept, were raised and investigated. The 1930s saw the rise of a number of doubts and confusions about the organizer, and the 1940s saw its abandonment by some and considerable modification by others. To trace the later developments in embryological theory, and to begin to understand the synthesis of disciplines which was underway, it will help to understand a few of the problems that workers encountered in following out the implications of the organizer concept.

Certainly one of the most immediate as well as typical problems emerging from Spemann's organizer concept was the nature of the organizer itself, the mechanism by which the dorsal lip, or any other inducer, exerted its influence. As Spemann pointed out, the evidence available in the 1920s argued strongly for a chemical agent emanating from the inducer to the induced tissue. In a flurry of activity embryologists engaged in a search to isolate and identify "the organizer substance." This search involved use of many techniques introduced from biochemistry and physiological chemistry and gave impetus to a burgeoning new field, biochemical embryology. Enthusiastic young workers such as Joseph Needham (1900-   ) at Cambridge, looked with particular interest on the organizer concept because it lent itself well to chemical analysis. As Gavin de Beer and Julian Huxley, two optimistic young embryologists remarked in 1934:

"It may be confidently expected that in time the physiological basis of the organiser's action will be discovered and accurately analyzed in physico-chemical terms."

Their enthusiasm for this mechanistic and reductive approach led Huxley and De Beer to quote with confidence that, based on the work of two other embryologists, Needham and Waddington (1933), the exact nature of the organizing influence was known: it was almost certainly a lipoid and probably a steroid.

Yet even as biochemists were pointing triumphantly at one or another

specific molecule, a number of other workers were turning up some disconcerting facts about the "organizer."

One of these was that heat-killed organizer tissue (from the dorsal lip)
could still induce ectodermal tissue to differentiate. This observation could
be explained by maintaining that the specific molecules of the inducer
substance were heat stable and thus not affected by temperature. However, it
was also shown that the inductive capacity of heat-killed or chemically killed
dorsal lip tissue had widely differing inductive capacities on different embryonic areas. Thus, for instance, heat-killed organizer tissue lost nearly all
its capacity to induce differentiation of mesodermal tissues, but retained
almost full ectodermal capacity. In view of this observation, it was difficult
to speak of one organizer substance, since it appeared that perhaps several
substances, each specific for one germ layer region, emanated from the dosal
lip. Still more disconcerting was the demonstration that many nonspecific
substances—inorganic chemicals, pieces of adult liver or spleen, steroids,
fatty acids, carbohydrates, changes in ionic concentration and pH, or even
dust from the floor—could all evoke organizerlike responses from developing embryos. It thus appeared that the organizer substance was either (1) not
a single substance, or (2) very general in its nature, the specificity of any
particular inductive process being dependent on other factors, about which
little or nothing was currently known.

In his early years, Spemann had indirectly suggested that the organizer
produced some specific chemical substance by which it exerted its effect. But
he was never wholly explicit on the matter and was, in fact, relatively
uninterested in biochemistry per se and its bearing on induction. Although
his ideas stimulated much biochemical work, Spemann himself remained
relatively vague on the subject of what the organizer really was. Although he
was not a vitalist, he did claim in his early years that the organizer possessed
"all those enigmatic peculiarities which are known to us only from living
organisms," by which he meant self-maintenance, organization, response to
stimuli, and so forth. In his later years, especially after biochemical research
had failed to identify any specific organizer substance, Spemann imbued the
organizer with almost psychic qualities, pointing out again and again that:

"The reaction of a germ fragment, endowed with the most diverse
potencies . . . is not a common chemical reaction, but . . . like all vital
processes [is] comparable . . . to nothing we know in such a degree as to those
vital processes of which we have the most intimate knowledge, viz., the
psychical ones."

The dorsal lip and the process of organization that it could direct was not outside the realm of known law, any more than, to Spemann, the ability of the human mind to "think" defied chemical and physical principles. But unlike the simplified mechanism of Roux and Driesch, Spemann's philosophy was more holistic.

This can be seen in Spemann's changing views on the function of the organizer *vis à vis* the organized tissue. Early on, Spemann had emphasized almost exclusively the remarkable capacity of the dorsal lip tissue to induce the formation of secondary embryos. But as time went on Spemann, and especially his students, recognized that induction and organization were not the functions of the dorsal lip tissue alone but also depended on the nature of the reactive tissues. In fact, as time went on, the discovery of reciprocity— that the induced tissue has reciprocal influence on the inducer—forced Spemann and others to recognize that differentiation was the function of a whole system, not of isolated parts such as dorsal lip tissues.

Spemann began to see that induction was not a simple mechanism, such as pulling a trigger, from which a whole sequence of events proceeded. Instead, the initial induction produced a reaction in some target tissues, which in turn influenced the future activity of the inducer. It was not a one-way military hierarchy, but a multidirectional system of many interrelating effects. Historically what is significant about Spemann's thought process is that he first made a distinction between the inducer and the reactive system, showing clearly that two such components existed as part of any differentiative process. Had he left the matter there, however, he would have provided only a kind of mechanistic analogy similar to those proposed by Roux, Weismann, or others. By emphasizing the interaction of the two components, each changing the other, he moved from mechanistic to holistic (or dialectical) materialism and thus pushed the study of differentiation onto a new plane. Like the holistic physiology of Cannon or Henderson, the interactive systems of Spemann were concerned with the complex of events surrounding particular phenomena. Spemann felt this complexity could not be understood simply by reducing the biological phenomenon to individual chemical and physical reactions. Something else was involved; a level of organization—tissue interdependency—that defied detection if approached in a purely mechanistic and reductionist way.

Yet it is not difficult to understand how Spemann's views were misinterpreted and how embryology took an ambiguous turn as a result of his work. On the one hand Spemann himself was never very clear on what he thought the organizer did or on what sort of material basis he conceived induction to operate. He was not a vitalist and therefore assumed that some material influence must emanate from dorsal lip tissue (or any other inducer). In fact, his reference to the psychical quality of organizers could only lead to confusion and misinterpretation among many of his followers. Moreover, Spemann claimed that he never constructed an "organizer theory" but only coined a term to describe the phenomenon of induction of a secondary embryo by dorsal lip tissue:

"I coined the term 'organizer' to describe some new and very remarkable facts which I ran across during my experiments. I have, however, from the beginning considered this conception a preliminary one and have, more than once and in a formal manner, characterized it as such. I have nothing to do with any attempts to make it the foundation of a theory."

Historically speaking, of course, Spemann did produce an "organizer theory," although he may not have intended to and did, in fact, repudiate it as a general principle. The idea itself was too catchy and too exciting to escape being generalized, and Spemann did not avoid generalizations altogether himself. So confusing and chaotic did many of the facts of development appear, and so little was there in the way of any general paradigm to account for the age-old, yet burning, issue of differentiation, that the organizer concept was taken up with considerable enthusiasm and often with a less than critical eye. Methodologically and intellectually Spemann maintained his reservations. Many others did not.

The work of Spemann and his school took two different directions during the late 1920s and early 1930s. One, which we have already discussed in some detail, was that of biochemical embryology—largely the search for the chemical mechanism of induction. The second was toward a more careful analysis of developmental systems on the gross (tissue) level, leading ultimately to the foundation of embryological field theory.

Despite his own lack of interest in biochemical processes, it is easy to understand why Spemann's work on induction in general, and the dorsal lip in particular, should inspire the search for *the* organizer substance. At any period in modern biological history it would have been a natural question. By what means does the inducer stimulate development in inducible tissue? But especially in the 1920s, the heyday of the mechanistic approach to life, it was inevitable that the organizer would be assumed to have a single chemical and physical basis that should be discoverable by the appropriate experimental analysis. Under the influence of thinking such as Loeb's, embryologists often translated the term "organizer" into "substance" with specific chemical properties. The rampant tendency to find *the* single explanation for any phenomenon caused some embryologists at least to do for Spemann's concept of the organizer what Loeb did for his own theory of instinctual behavior: endow it with more chemical and physical specificity than the evidence could support. It is important to remember that it was more the atmosphere in which the idea was launched than Spemann himself that was responsible for the highly mechanistic and oversimplified search for a chemical organizer in the 1920s and 1930s. Those who later claimed that Spemann's Nobel Prize should be revoked because organization did not seem to be a specific phenomenon describable in physical and chemical terms were showing their own mechanistic bias. Such an explicit connection was not a component (to any degree) in Spemann's original idea, and it is to his credit that he always resisted the tendency to think in such simplistic terms.

The other direction of Spemann's influence arose somewhat later in time—the 1930s and 1940s—as an outgrowth of transplantation studies between embryos of different ages and species. These experiments led many workers to focus their attention on the complex interaction between inducer and induced tissues. Taking their cue from Spemann's own writings, young developmental biologists such as Paul Weiss (1898-) came to regard "organizer" and "organized" tissues as parts of a "morphogenetic field" and, from this point of view, sought to develop a field theory for biology not unlike the one that was beginning to emerge about the same time in the physical sciences. The field theory in embryology was an attempt to devise some means of comprehending the plastic, multidimensional interactions that seemed to characterize, both in time and space, the process of self-regulated growth and differentiation typical of embryonic development.

Early in their work Spemann and his students had noted that embryos seem to be organized into regions, or geographic areas, that function as an integrated whole to produce particular structures or groups of structures. These regions, characterized by considerable flexibility and capacity for self-regulation, are what Spemann, Weiss, and others came to regard as "fields" within which certain forces work to produce a particular unified set of structures. The field theory was in many ways similar to Driesch's 1891 suggestion that the whole embryo is a "harmonious equi-potential system," adjusting to and repairing itself from the effects of external modifications.

A simple case will illustrate how field theory was applied to the differentiative process. Recall that Spemann and Mangold had shown in the early 1920s that the eye lens develops from ectodern that comes into direct contact with the underlying optic cup tissue. If presumptive lens tissue (ectoderm that would become a lens if left in position) is removed from an early amphibian embryo (at a stage past the gastrula but considerably before actual eye structures would normally appear), the lens still develops! The surrounding ectodermal tissue moves in to close the wound and, in so doing, comes in contact with the underlying optic cup. The conditions again exist for induction, and the lens is formed. Ectodermal cells that would never have formed a lens under ordinary circumstances do so after the operation because they are part of the "lens field." The field is the region within which such adjustment responses are possible. The lens field is much larger than the group of cells that normally form the lens itself; it extends to all the ectodermal tissue that can become involved in lens formation if the original cells are removed. Fields are regions or areas of tissue that can readjust within themselves to changing conditions. According to field theory, the developing embryo is seen to consist of numerous such fields, many times overlapping each other and changing with time as the embryo grows. The field concept emphasized the enormous flexibility of the embryo to readjust itself continually, something that was difficult to explain on simple mechanistic and chemical models.

In helping embryologists understand such complex phenomena, the field theory has been highly influential. Many embryologists today claim that despite its vagueness, the field approach has been one of the most important influences in their conception of the complex interrelationships involved in differentiation. The field theory is far removed from the concept of rigid, inductive influences passing down a militarylike hierarchy in the embryo. It emphasizes the integrity of certain regions in carrying out their differentiation process. Yet, despite its value, which still persists, in providing a picture of the complexities of embryonic development, the field theory gave vitually no specific direction to research on the problem of exactly how differentiation is triggered or maintained. It was the logical and perhaps final extension to which the holistic side of Spemann's work could be carried.

Spemann's approach to differentiation was carried out mostly on the supracellular level, which marked a considerable departure from the early stages of the *Entwicklungsmechanik* tradition. However, developmental mechanics as it evolved from the work of Roux, Driesch, the young T.H. Morgan, Jacques Loeb, E.B. Wilson, and others was largely aimed at the cellular level. With their strong mechanistic predilection, it was only natural that these workers would have focused the methods of *Entwicklungsmechanik* on intracellular events. The theories of Roux, Driesch, and Weismann were theories of how hereditary factors were divided between cells during embryogenesis. Similarly, the early experiments of Loeb, Wilson, Morgan or Conklin were concerned with studies of eggs, nuclei, or chromosomes. Cells were recognized to be parts of tissues, and tissues parts of organs, but the questions that were being asked were questions about what went on inside of cells. E.B. Wilson's monumental *The Cell in Development and Hereditary* (first edition published in 1896, second edition in 1900) is an eloquent testament to both the interest prevalent, and the high level of achievement reached in the study of cells (cytology) by the early 1900s. Wilson's book, more than almost any other, epitomized the importance that biologists were beginning to attach to the cell as the fundamental unit of physiology, heredity, development, and even evolution. The *Entwicklungsmechanik* tradition arose in an environment where cells were considered the basic unit in which all living phenomena could be localized. The early experimental embryologists were cell biologists, and their most popular cells for study were the egg (fertilized or unfertilized) and the early blastomeres, in which they thought the differentiative process was already laid out and, in some sense, determined.

Spemann's work picked up the method of experimentation from *Entwicklungsmechanik* and applied it to the supracellular level—the level of tissues and organs. Although he did some early experiments of his own on the fertilized egg and the initial stages of embryonic growth, his major work was with older embryos and the interaction between groups of cells and tissues. He applied the classical methods of experimental embryology, such as removal of an early stage, to determine what effects, if any, result in the later embryo, and invented new methods such as transplanation between embryos to investigate the differentiation process. The questions that Spemann asked took embryology away from the single cell where techniques of biochemical analysis were too imperfect to provide any answers about differentiation and brought it to the level of tissue interactions where experimental methods could be devised and applied with considerable success. Spemann, of course, believed that differentiation took place ultimately on the cellular level, but he knew that no answers about the larger questions of how the process was triggered, or how it proceeded with such regularity and precision could be obtained by studying only the fertilized egg, or the early blastomere. Like Morgan and his group, who admitted that questions of the function of genes were important but continued to ask other kinds of questions because they could be more readily answered, Spemann knew what to ask of the embryo and how to devise techniques to gain an answer.

To return to our initial point, however, we have to ask what was joined to what in the convergence of disciplines supposedly centering around embryology and genetics. The fusion was initially more conceptual than experimental, but with time the experimental aspect became more possible as well. Knowledge of classical Mendelian genetics dispensed with the need among embryologists for Roux's mosaic theory, emphasizing that the problem of differentiation was one of differential activity of genes. Physiological geneticists such as Goldschmidt actively sought to make explicit the link between the gene and the expressed character. Much was speculation, but the realization was nonetheless common among many biologists that gene action had a biochemical basis that, if understood, would provide a key to the mystery of differentiation. So vivid was the concept of a present or impending synthesis that T.H. Morgan wrote a book in 1934 entitled *Embryology and Genetics* dedicated to showing that "the story of genetics has become so interwoven with that of experimental embryology that the two can now to some extent be told as a single story." After outlining the major events in maturation of egg and sperm and chromosome structure, Morgan posed the central problem of the day: how is it that different genes appear to act at different times during development, so that some cells differentiate in one direction, and other cells in another? His answer was only a speculation—a return to an old idea that initial regional differences in egg cytoplasm may provide the differential trigger for gene action. In the end Morgan was unable to make the synthesis explicit, since there were then no techniques to study such problems on the level of the cell or the molecule where they had to be studied. A colleague had expressed to Morgan his disappointment that *Embryology and Genetics* did not make more explicit and concrete the relationship between Mendelian genetics and the mechanism of differentiation. Morgan is reported to have remarked: "I did just what the title of his book suggests. I discussed embryology *and* I discussed genetics!" Such was the degree of synthesis on the experimental level that was possible by 1934. On the conceptual level, however, the synthesis was becoming clearer and more imminent as time went on.

Ultimately, answers to the problem of differentiation were to be obtained on the cellular and subcellular level—particularly by the 1950s and 1960s. There are two main reasons for a return to the cellular level of investigation in these recent years. One is that work on the supracellular level—including the organizer and field theories, as valuable as they had been in some ways, did not provide direct answers or even directions to further research. The other is that the development of new techniques such as radioactive tracers, electrophoresis, and chromatography (for isolating and identifying chemical components of cells) and microsurgery (for transplanting nuclei or cytoplasmic fragments) made it possible to carry out experiments on single cells (the egg or embryonic cells) about which Spemann and Morgan could only speculate. With such techniques as microsurgery, for example, Robert Briggs and Thomas King in the United States and John Gurdon in England, in the middle and late 1950s tried to determine whether nuclei in embryonic cells were reversibly or irreversibly restricted in their potentialities during normal differentiation. Their basic method involved removing the nucleus from an unfertilized egg of one amphibian species and replacing it with a fertilized nucleus, or a nucleus from some type of adult cell of the same or different species, and observing the state of differentiation to which the new embryo could grow and the kinds of abnormalities if any, that occurred. In addition Gurdon and others began to look at molecular changes—specifically in ribonucleic acid—during early development to determine how the differentiation process might be related to gene activity.

The return to cellular-based question in the 1950s and 1960s was not a return to the oversimplified mechanism of the early *Entwicklungsmechanik* days. The role of the cell in development had now become visible against a backdrop of highly complex tissue interactions that indicated that differentiation was not the result of a single mechanism operating at all times in all parts of the embryo. The full synthesis of embryology with genetics could not occur until the revolution in molecular biology in the 1950s and 1960s provided new techniques and new concepts at the biochemical level for understanding how genes have their effects. Certainly a major innovation of the 1950s and 1960s was the use of highly simple organisms—in many cases bacteria and other protists whose single-cell nature meant that the problems of development were largely changes in biochemical quantities (the appearance or disappearance of certain proteins, or ribonucleic acid types). New techniques made it possible to investigate such developmental changes accurately and quantitatively on the cellular level.

PART II

The Evolutionary Synthesis

The First Steps Toward Population Genetics

During the same period that embryologists were trying to bring their discipline into line with modern methods of biochemistry and cytology, a young group of biologists was trying to relate evolutionary theory of the neo-Darwinian school to the new concepts of Mendelian genetics and classical natural history. In effecting the evolutionary synthesis evolutionists were more fortunate than embryologists in a number of ways. The advances in Mendelian genetics had considerably more direct and obvious application to evolution than they seemed to have in embryology. The new methods that evolutionists needed to apply Mendelian genetics to Darwinian theory (mathematics) were more readily developed in the 1930s than the biochemical and other techniques needed by embryologists to relate Mendelism and cell physiology to induction or organizer phenomena.

The unification of evolutionary and genetic theory did have to overcome a number of obstacles, however, that made the task piecemeal and hesitant. Many of these were specific; for example, the residual antagonism from the previous century between field naturalists and laboratory experimenters, or the continuing debate between biometricians and Mendelians over whether continuous or discontinuous variations were the raw material of evolution. As large as these obstacles loomed, there was one still larger and more fundamental. This was the tendency on the part of naturalists and laboratory experimenters alike to think of species in terms of individual types and consequently to view evolution as change in individuals instead of as change in populations.

Crucial to the development of a populational approach to evolution was the fusion of genetic concepts with the field data from natural history. This was made possible partly by a line of thought that reduced populations to their composite genes, and by another that found mathematical means to predict how gene frequencies would change under particular environmental conditions and selection pressures. Both of these approaches were aided by attempts to formulate new concepts of systematics—of what a species really was—that were becoming prominent (at least among some workers) around the turn of the century. Although the new species concept did not reach a full or very widespread development until well into the 1940s, the synthesis between evolution and genetics had taken its first major steps several decades earlier.

We have seen that the Darwinian theory in 1900 was incomplete in its understanding of a number of points: (1) the origin of variations; (2) what type of variations were heritable and what type were not; (3) the role of isolation; (4) the role of selection; and (5) the nature of species—were they real units in nature, or only arbitrary subdivisions made by man? The stimulus to fuse evolutionary theory with other branches of biology in the 1920s and 1930s was the result of a strong desire on the part of many workers to fill in these gaps and to make the theory of natural selection logically complete. Even among the Darwinians, many points could not be made except in the most qualitative terms. Convinced Darwinians had no way of demonstrating the necessity of evolution arising out of the mechanism of variation and selection. They knew it was right, but they were unable to convince those who *felt* it was not.

By itself the biometrical movement could never have led to the formulation of a synthetic view of evolution. Because of their adherence to the old laws of filial regression or ancestral inheritance, biometricians were unable to understand how variations could persist in a population without being "swamped"; and, because they rejected Mendelism as "particulate" and discontinuous inheritance, they closed the door to any further understanding of the origin or transmission of variations. Although Mendelian genetics lent itself very well to statistical treatment on a population level, the biometricians were unable to make a transition between the Mendelian gene and the Darwinian variation. What the biometricians did contribute was a *statistical manner* of thinking, a means of approaching hereditary problems with large *populations*. Despite everything, however, theirs was a *populational* approach to heredity. This outlook was picked up by many workers—both inside and outside the strict biometrical camp. In England it conditioned the biological thinking of a whole younger generation of workers, such as J.B.S. Haldane or R.A. Fisher, who later became major contributors to the synthesis of evolution, natural history, and genetics.

Some of the first applications of statistical thinking to simple population problems grew out of Mendel's original work. In his 1866 paper Mendel pointed out that one of the automatic consequences of segregation in self-fertilizing organisms (such as peas) would be an orderly and predictable decrease of the proportion of impure or mixed forms (called heterozygotes in later terminology) in successive generations from a cross between two initial pure forms (called later homozygotes). Thus, according to this idea, inbreeding (brother-sisters crosses) would result in the gradual reversion of the offspring populations to the original (homozygote) parent types. The key feature of this generalization was that it specified breeding conditions: when inbreeding (nonrandom mating) is the rule, the frequency of heterozygotes will be expected to decline.

Shortly after the rediscovery of Mendel, however, several other workers saw the opposite side of the coin. If the proportion of heterozygotes should decline during inbreeding, then it should remain constant if there is crossbreeding, that is, if mating is random. The idea was already in the air that an "equilibrium" would be established in any population where mating was not restricted. This generalization received a full mathematical treatment by the Cambridge mathematician G.H. Hardy (1877-1947) in 1908, and quite independently by a Stuttgart physician, Wilhelm Weinberg (1862-1937) in the same year.

What Hardy, Weinberg, and others pointed out was that the 1:2:1 Mendelian ratio (or any other ratio for that matter) would be stable from one generation to the next under conditions of random breeding and that this could be expressed as the expansion of the algebraic bionomial.[1] Thus,

---

[1] Mathematically, this could be written as $p^2 + 2pq + q^2 = 1.0$. Where p = the frequency of dominant allele (e.g., A) and q = the frequency of the recessive allele (e.g., a) in the population as a whole. p^2 can be read as AA, and pq as Aa.

whatever the original frequencies of two (or more) alleles happened to be in a population at the outset, these frequencies would be maintained throughout successive generations. The actual 1:2:1 proportion represented only one of an infinite number of possible gene frequencies in a population, since frequency was usually determined for any population at a given time by a variety of circumstances, some of which could be purely accidental.

While not attracting a great deal of attention in the early 1900s, the Hardy-Weinberg concept was the quantitative expression of a more general understanding that was becoming apparent in a qualitative way among many geneticists in the first two decades of the century: that gene frequencies, or proportions, would remain constant of their own accord from one generation to next unless acted on by outside influences, such as selection or nonrandom mating. More important is the fact that to reach such a conclusion it was necessary to begin thinking of heredity in terms of populations instead of in terms of individuals. The biometrical movement had aided this realization by providing a climate in which statistical thinking about heredity had become an acceptable, if not a frequently practiced, approach among many biologists. Biometricians and those following the Hardy-Weinberg concept were alike in applying statistical and populational thinking to heredity questions. However, they differed in their use of Mendelian genetics and data from natural populations. The biometricians studied natural (and human) populations, but they were non-Mendelians. Hardy, Weinberg, and others were Mendelians, but they did not study natural populations.

Mendelian and the Study of Natural Populations

When H.J. Muller travelled to the Soviet Union in 1922, he wanted to learn about the new Socialist experiment being carried out there. He took with him not only a deep-rooted sense of the validity of the Communist movement, but also a number of pure-bred *Drosophila* stocks that were until then unavailable in the Soviet Union. However, the legacy that Muller left behind after several months' stay was more than a few bottles of flies with detailed pedigrees. He communicated to his Russian colleagues his excitement about the new Mendelian chromosome theory, which he had helped to create. He also must have communicated his own concern with the relations between the new genetics and evolution; of all those in the Morgan group, Muller was the one who had, from the beginning, been most intrigued with heredity and its relation to evolutionary theory.

Muller's message did not fall on deaf ears. Russia had had a long tradition of field work in natural history. With no strong biometrical, neo-

Lamarckian, or even Mendelian school prior to the Revolution, there was little of the sort of opposition to the aims or methods of natural history that had emerged in the West around the turn of the century. Thus the young workers to whom Muller talked were not inclined, as in the United States, England, or Germany, to reject the study of evolution as nonrigorous or speculative science. At the same time, the Russians were immediately captivated by the methods, techniques, and conclusions of the Morgan School, and perhaps as much through Muller's personal influence as anything else, saw no incompatibility between Mendelism and Darwinism. In no other country at that time (the 1920s) did leading biologists accept as a rule both of these fields so completely.[2]

Among the most receptive of Russia's biologists to Muller's stimulus was Sergei S. Chetverikov (1880-1959) and his students, D.D. Romashov (1906?-1965) and N.P. Dubinin (1907-   ). Between 1920 and the middle 1930s their work laid the foundation for applying Mendelian genetics to the hereditary properties of natural populations over numerous generations. They developed simple mathematical and experimental models whose validity they tested with field or laboratory populations.

Trained as a systematist, Chetverikov developed early in his life an interest in the large-scale problems of evolution toward which taxonomy supposedly (but all too rarely) leads. His reading of the evolutionary literature brought him into contact with the work of numerous neo-Darwinians, and in particular the writing of John T. Gulick (1832-1923), the British missionary and naturalist. Gulick had studied land snail populations in Hawaii and developed an evolutionary concept of geographic isolation that was of much interest to Chetverikov (although now known to be invalid). From these readings and from his own studies of butterfly populations, Cheterikov became interested in population dynamics in insects. Particularly intriguing to him was the evolutionary significance of "population waves," drastic increases and decreases in population numbers, a subject on which he wrote while still a teacher in the prerevolutionary Moscow University for Women. After Muller's visit, as a faculty member at the newly constituted Moscow University, Chetverikov became interested in the questions of the origin of variation in natural populations and the relation between population size and evolutionary potential.

[2] In the United States Mendelian genetics was well accepted, but the aims and methods of natural history were in disrepute. In England, natural history was considered valid but, by the 1920s, through Bateson's recalcitrance on chromosomes, the Mendelian-chromosome theory had dwindled in extent and enthusiasm. In France and Germany both ideas were viewed with considerable skepticism.

From his obsevations of natural populations, Chetverikov concluded that all populations display a large amount of unseen or "cryptic variability," which could have enormous consequences for evolution. These variations were recessive (hence not readily seen) and in many cases lethal (would cause death if occurring in the pure of homozygous form). The proportion (frequency) of these recessive genes would also be expected to remain constant from one generation to another, in agreement with Hardy-Weinberg expectations, unless the homozygotes were selected against. With his students, Chetverikov tried to test this reasoning during the summer of 1925 with the genetic analysis of a natural population of *Drosophila* (from the environs of Moscow). By crossbreeding the wild flies with flies of known genetic constitution (from Muller's original stocks) and with other tests, Chetverikov showed that out of a population of 239 flies, there were detectable at least 32 masked recessive traits (and perhaps many more that simply had not been detected).

In a 1926 paper, titled (in English translation) "On Certain Features of the Evolutionary Process from the Viewpoint of Modern Genetics," Chetverikov calculated from simple equations that he devised, for a variety of gene frequencies the probabilities that recessive variations would appear as homozygotes. He concluded that the larger the population, the less likely that cryptic variability would be shown, and the smaller the population, the more likely. For this reason, evolution would be expected to occur more rapidly in small than in large populations. Like Gulick, whose work had influenced him so strongly, Chetverikov came to emphasize the importance of isolation, by producing small population groups, as a critical factor in evolution.

Two of Chetverikov's students, D.D. Romashov and N.P. Dubinin, set out to test these conclusions by applying the principles of probability (which Chetverikov had only barely hinted at) to models of populations under various conditions of size, amount of variability, and different gene frequencies. With rare ingenuity Romashov and Dubinin introduced a model system known as "the urn experiment," one of the most useful and influential applications of statistical principles to the study of population genetics developed in the past 50 years. Using a large urn or jar into which markers (marbles, usually) of two or more different colors could be placed, each color representing one of several possible forms for any particular gene, they withdrew two markers at a time, each marker representing one of the two genes that the offspring will inherit from each parent. They found that the proportions of various genotypes will always bear a distinct relation to the frequency of each type of gene in the preceding generation's total gamete output. The urn model not only allowed a rigorous experimental test of equilibrium concepts, which was simple enough, but it also allowed Romashov and Dubinin to test directly the effects of various initial conditions (such as positive or negative selection, or nonrandom mating) on gene frequencies in subsequent generations. The urn became a model of the reservoir of the population's genes—its so-called "gene pool"—in which all combinations of genes could be realized. Which ones *were* realized depended on laws of probability, on which could be superimposed a host of special environmental conditions.

With the urn experiments, Romashov and Dubinin corroborated a number of Chetverikov's conclusions, including the seeming advantage of small populations for effecting significant changes. Although other theoretical considerations, emphasized clearly by R.A. Fisher several years later, produced the opposite conclusion, what is important is that the Russian workers brought the quantitative and rigorous approach of the Mendelian school to bear on the field studies of natural history. Furthermore, by their emphasis on the "gene pool," they treated populations not simply as collections of individual organisms, but statistically as "collections of genes." Galton, Pearson, and the biometricians had been forced to apply their statistics only to phenotypes of a population, never being able to separate hereditary from environmental effects, and thus being unable to determine whether real evolutionary changes had occurred or not. Although not as mathematically sophisticated as the biometricians or later population geneticists, Chetverikov and his collaborators opened up an entirely new perspective to the study of evolution.

The Russian school developed at a considerable pace in the 1920s and early 1930s. In addition to Chetverikov, Romashov, and Dubinin, other geneticists, evolutionists and, to a more limited extent, statisticians were all involved in working out the genetic structure of various *Drosophila* populations on the basis of statistical sampling and model building. The development of population biology as an especially Russian field at this time was partly a result of the impetus given to genetic work by Muller's visit with his special interests in mutation and evolution. It is also partly a result of the determination of the new Soviet regime to foster science, and particularly biology in the immediate postrevolutionary years. For example, when Kolstov walked with Lenin during the 1920 Leningrad famine, Lenin said "The famine to prevent is the next one and the time to begin is now." As a result, emergency funds were spent partly to build an Institute of Applied Botany, and later to build the Institute of Animal Genetics in Moscow. In those years the Soviets had high regard to the new biology, that is, for genetics and its relations to evolutionary theory.

Particularly ironic, therefore, was the later denunciation of Mendelian genetics and the chromosome theory of heredity by T.D. Lysenko (1898-   ) and his followers, with the full ideological support of Stalin and the Central Committee of the Russian Communist Party. Lysenko became the chief exponent of a revived interest in neo-Lamarckian inheritance (the influence of the environment on heredity) that he used as the basis for an extensive and ultimately valid agricultural breeding program. Lysenko claimed that neo-Lamarckism, because of its emphasis on environmental conditioning as a determining feature in an organism's development (as opposed to heredity), was more amenable to the philosophy of materialism than the "idealistic" theory of the Morgan School. Lysenko gained Stalin's confidence and effectively eliminated or drove underground much Mendelian research in the Soviet Union between the middle and late 1940s and the middle 1950s.

There is no doubt that the actual doctrines of Lysenko, as they related to concepts about species, selection, and the influence of the environment on the germ plasm, were based on considerable ignorance, little data, and much speculation. These ideas had been discredited outside of the Soviet Union since the middle or late 1930s. This makes all the more curious enormous influence and position of Lysenko in the 1930s and 1940s. At least a significant part of the answer can be found in Russian historical circumstances during the period preceding and including Lysenko's rise to fame. The hardships of the 1917 revolution, on top of World War I, had left Russia in a chaotic and economically insecure situation. Particularly difficult in the early 1920s was the problem of hunger and famine. As a result of the social upheaval surrounding the revolution, agricultural production had been slowed. To make matters worse, many landowners and aristocrats, in fleeing the country, had burned their crops and slaughtered their livestock instead of allowing them to be confiscated and used by the masses. Furthermore, a great drought from 1921 to 1922 brought hundreds of thousands to the brink of starvation, making it apparent to Lenin and other Soviet leaders that agricultural planning had to be reorganized immediately lest famine, a never too distant threat in past Russian history, continue to rear its head in ensuing decades.

For these reasons considerable emphasis was placed in Lenin's time on scientific agriculture and rational programs for improving agricultural output. In the late 1920s Michurin and his follower Lysenko provided a program that, rejecting the slow, painful methods of orthodox breeders, namely selection, as a means of improving a stock, promised quick results by exposing organisms to the direct action of the environment. Academic biologists recognized the difficulty with this proposal and suggested that agricultural aims should be pursued through the avenues of pure research exemplified by Mendelian genetics. While the academic biologists were right in the long run, it was difficult for anyone to see at the time how deploying immense resources for studying eye color or wing shape in the tiny fruit fly would ever help solve the pressing problem of recurrent famine. Lysenko and the government leaders who followed him made the error of seeking a royal road to crop improvement. The academic biologists, on the other hand, erred in keeping their work too aloof from the very real and human problems that the society at that time faced. It is important to remember that Russian leaders, faced with so many problems on all sides, could hardly have failed in good conscience at least to give the Michurin and Lysenko programs a try. Where they made the most serious error was in not analyzing their own results carefully enough and not seeking out the advice of either peasants, who actually worked with the Lysenko methods in the field, or academic geneticists, who knew about the principles of hereditary and selection.

The gains that Chetverikov, Dubinin, Romashov, and other Russian workers had made in population genetics in the 1920s were not lost to posterity, however. The ideas were available in the literature, and Theodosius Dobzhansky (1900-   ), a student of Chetverikov, brought the Russian populational approach with him when he emigrated from Russia to Morgan's lab at Columbia in 1927. Dobzhansky's interest and influence, to which we return later, was not the immediate starting place for the synthetic theory of evolution, although his work became an important component of that synthesis in the 1940s. The next step was taken by a group of mathematical geneticists in England, and we now turn to their work.

The Synthetic Theory of Natural Selection

The biometricial school of Galton and Pearson had brought mathematical and statistical thinking to bear on the problem of inheritance and had thus bridged the gap between the heredity of individuals and that of populations. The Russian School of Chetverikov had brought Mendelian genetics to bear on the problem of heredity in natural populations. These two separate syntheses were themselves brought together principally through the work of R.A. Fisher (1890-1962) and J.B.S. Haldane (1892-1964) in England, and Sewall Wright (1889-   ) in the United States. These men increased the mathematization of population genetics by introducing new statistical concepts for dealing with factors such as inbreeding, variance, linkage, and multiple interactions. They thus were able to treat the changes in gene frequencies from one generation to another more rigorously than Chetverikov and his students had been able to do. The reason for this was simple. Like Henderson's multifactor equilibrium systems, population dynamics involve the interaction of a number of components. In both cases new methods had to be devised to treat the interaction of these many components simultaneously. For Henderson it was the nomogram, while for the population geneticists it was statistical and mathematical techniques that they inherited from traditional statistics or invented themselves to meet the very special situations they encountered. Through these new approaches, a whole range of phenomena, which otherwise stood separately from each other, was shown to be interrelated.

Haldane, Fisher, and Wright worked somewhat independently and along slightly different lines. All three were especially interested in showing quantitatively how natural selection would operate under a number of environmental conditions, but they came at the problem from different angles and with different skills. Fisher was the most mathematically sophisticated and ingenious, having studied both statistics and biometry at Gonville and Caius College, Cambridge (1910-1913). Steeped in the biometrical tradition, Fisher had the least formal or systematic knowledge of biology. His contribution to the synthetic theory was, not surprisingly, the most theoretical and abstract.

J.B.S. Haldane was, from the outset, more biologically oriented and trained, studying mathematics, biochemistry, genetics, and physiology with equal ease. The son of the Oxford physiologist John Scott Haldane (1860-1936), he was early interested in biochemistry, serving as a Reader in Biochemistry at Cambridge and later as a professor of genetics and biometry at University College, London. Haldane's work in population genetics combined a mathematical approach to populations with an interest in specific problems of genetics, such as the nature and origin of dominance or the factors that influence gene fitness.

Sewall Wright was the most traditionally trained of the three population geneticists. Working under W.E. Castle at Harvard, he had studied originally the genetics of coat coloration and its relationship to pigment biochemistry in mammals. Castle had introduced him to both the importance of Mendelian genetics and the genetic effects of selection on the phenotypes of populations. From the beginning Wright was brought to see the compatibility between Mendelian and Darwinian theory. Knowledge of both was useful to him as a senior animal breeder (1915-1925) for the U.S. Department of Agriculture. After leaving government service, he taught at the Universities of Chicago and Wisconsin. With no formal training to speak of in mathematics, what Wright learned of this field he virtually taught himself. His major emphasis in population genetics was detailing in a quantitative way the factors that disturbed genetic equilibrium. Thus he came especially to focus on the question of population structure (size and geographic development) and its influence on evolutionary change.

It is somewhat beyond the scope of this book to trace out the historical intricacies involved in who among Haldane, Fisher, and Wright influenced whom, who got which ideas first, and so on. It will be more satisfactory to treat the synthetic theory of natural selection as a totality, showing simply the different views that each man brought to it, regardless of chronology. It is only necessary to keep in mind that the three did work to some extent (at least in the middle and late 1920s) independently and that, on many issues, they disagreed with some of each other's conclusions. However, for our purpose, we will treat their work as a unified whole.

Fisher's *The Genetical Theory of Natural Selection*, first published in 1930, encompasses work he had been carrying out since 1918. This book is viewed by most historians as laying the foundation for population genetics and its relations to evolutionary theory. The title is significant and, from it, we can infer what Fisher's aims as well as self-imposed limitations were. He purposefully chose to relate genetics to the process of *selection*, not to the evolutionary process as a whole. He eliminated from his model building consideration of such variables as migration, isolation, genetic recombination, and gene interaction. Consideration of such factors would only have complicated the basic task immensely: to develop a rigorous mathematical model to show how Darwinian fitness was mathematically related to population growth rate, and the change in gene frequencies within populations.

Fisher pointed out that the vital statistics of an organism in relation to its environment provide a means of determining the relative growth rate of a population (he called this the Malthusian parameter of population increase). The Malthusian parameter of measures the reproductive values of individuals at all ages or stages of their life history and thus would be expected to be different for each genotype. The Malthusian parameter provides an index of the fitness of each genotype to survive under specific environmental conditions. Applying sampling and statistical techniques developed in biometrical and actuarial practices, Fisher was able to show how data from an existing population could supply information about the reproductive capacity of the population at the point in time when the sampling was made, or at other, hypothetical points in time where particular biological and environmental parameters were specified. He then went on to define what he called *genetic variance*, a quantitative expression of the amount of variation any particular genetic trait could show in a population, separate from environmentally caused fluctuations. Genetic variance is determined for any population by sampling the population under a variety of environmental conditions (different season, times in the reproductive cycle, etc.). Any single measurement from a population will, of course, be an expression not only of gene variability but also of genetic recombination and environmental influence. But a whole range of samples, from many environmental circumstances, makes it possible, by statistical analysis of the variance, to separate true genetic from environmentally induced variability. This was a key step, because at last it allowed Fisher to analyze the effects of selection on Mendelian variations and provided a model by which the effects of numerous variables, including amount of variation or degree of selection, could be predicted quantitatively. Through this analysis Fisher was able to deal exclusively, yet legitimately, with genetically as contrasted to genetically plus environmentally induced variations.

Fisher then related his analysis of genetic variance to the process of natural selection. The first component of this model was genetic variation arising almost wholly, Fisher thought, from mutation. Drawing on H.J. Muller's studies during the 1920s, Fisher pointed out that mutations from one form of a gene to another occurred at specific rates, providing a quantitative measure of how rapidly new variations are introduced into a population. The second component of Fisher's model was natural selection, the factors in the environment that determined the possibilities of a gene or a group of genes being passed on (and in what quantities) to the next generation (the so-called fitness of the gene, or group of genes). There were two measures of fitness that could be applied to natural populations. One was applicable to two or more species of populations compared to each other and was measured as a simple increase in size of one population over another. In the particular environment or environments under consideration, the fittest group is the one that shows a relative increase in population size. The other measure of fitness applied to particular individuals or particular genes within a single population and was calculated as change in gene frequency over successive generations. Increased fitness for a gene was reflected by an increase in its frequency.

Fisher's models were highly simplified, but therein lay their importance. He carefully separated out the key strands in the fabric of populational variation. He showed that the observed range of variability in a population was the result of three separate factors: genetic, environmental, and the observer's sampling errors. By providing means of determining the limits of each, through his analysis of genetic variance, Fisher brought the tools of biometry directly to bear on problems of Mendelian genetics. By directing his interest toward showing how variability of a simple sort affected the fitness of the gene complex making up any organism at any time, he laid the foundation for understanding how natural selection alters gene frequencies. It was the task of other workers, especially Haldane and Wright, to investigate the further factors involved in the evolutionary process as a whole. Fisher was the first to admit that his mathematical models did not attempt to account for the entire evolutionary process. It was, however, a rigorous mathematical treatment of several prime factors. And, without his careful separation of genetic variation from environmental influences and problems of sampling error, it would have been impossible for workers in succeeding decades to treat rigorously the more complex conditions found to operate in natural populations.

Although he had begun his work before Fisher's *Genetical Theory* was published, Haldane was familiar with much of Fisher's earlier work and was indebted to him for many of the statistical tools that the latter had developed. Haldane was interested in treating mathematically a variety of factors that would influence the fitness of genes as Fisher described it. These factors included complete or incomplete dominance, random versus selective mating, and self-fertilization (where generational bility of generational overlap) versus cross-fertilization (where generational overlap can occur). Haldane expanded Fisher's mathematical model by studying a number of specific conditions that Fisher, for the sake of simplicity and clarity, had purposefully set aside. Yet his treatment, like that of Fisher, was also largely biometrical and theoretical. Both men were concerned less with data from real populations than with constructing models of populations under specified conditions. To both Fisher and Haldane, populations were abstractions to be described in terms of the frequency of particular genotypes.

Somewhat different from this was the work of Sewall Wright. Coming onto the population scene later than Haldane or Fisher, he profited much from their initial work and personally testified to the importance of Haldane's thinking for his own growing interest in population genetics. Having been a student of Castle's in the early teens, Wright acquired an interest in both Mendelian genetics and evolution by natural selection from a laboratory point of view. He was well prepared to understand the relationships between Mendelism and Darwinism and to view these relationships in the concrete terms of real populations.

Wright's basic approach to population genetics was to begin with a simple situation in which genetic equilibrium persisted from one generation to the next and then to investigate the factors that could upset that equilibrium. Equilibrium could be treated mathematically (as expansion of the bionomial; see footnote, p. 129), and the various factors that might affect equilibrium could be measured by the degree to which they could alter gene frequencies. The factors that Wright studied were those of which evolutionists and students of natural populations had long been aware: mutation, selection, and migration. Each of these, he showed, upset genetic equilibrium and shifted gene frequencies in one direction or another. More important, however, was Wright's view that these (and other) factors did not act on single genes but on groups of genes. Genes existed in a *milieu* that included not just the genes on the same chromosome, but the entire gene complement of the cell nucleus and, ultimately, of the whole organism. As he wrote in 1929: "the single gene's selection coefficient [i.e., its fitness] must be the function of the entire system of gene frequencies" of the population of which it is a part.

Wright saw the problem of evolution in terms of balancing factors within the whole population's gene pool. A population was, to Wright, like a living organism in which all the factors were closely interwoven and mutually interdependent. From this basic state of equilibrium various disturbing factors (migration, chromosomal aberrations, accidental catastrophies to the population) could cause significant and often irreversible changes in gene frequencies. These changes could be adaptive or nonadaptive, depending on the circumstances involved. Stable populations were normally well adapted to their environment—meaning their total genetic complement was at some sort of "adaptive peak," as Wright put it. Stability at this peak had been achieved only by selection acting on the total gene structure of the population. But no adaptive peak was permanent. Introduction of new variations into the population (by migration, mutation, recombination, etc.), or changes in the environment could increase selective pressures on the total gene complement (but, of course, more on certain specific genes than on others). Further shifts in the frequencies of certain genes would occur, and eventually a new adaptive peak would be reached.

Wright considered his equilibrium concept as analogous to physiological equilibria, although recognizing that in physiological systems equilibrium is usually preserved despite external changes, whereas in evolutionary systems one equilibrium gives way to another as external (environmental) or internal (genetic) conditions are altered. As we saw in Chapter IV, equilibrium ideas were finding their way into a number of areas of biology in the 1920s and 1930s. These ideas had in common not only the realization that stability can be maintained by the dynamic interaction of continually changing quantities, but also a new approach and methodology for describing and predicting changes in complex interacting systems. Perfect stability was not thought to be possible in either physiological or genetic systems: processes under such dynamic control continually oscillate around mean values. Despite the obvious differences between concepts of genetic and physiological equilibrium, they have certain common features that indicate the increasing attempts of biologists to deal with complex and interacting systems in a quantitative, rigorous, yet non-reductive way. Like Henderson, Fisher, Haldane, and Wright were forced to invent new techniques (mathematical) for dealing with such complex interactions. The equilibrium concept in population genetics allowed evolutionists to see quantitatively that adaptation is produced by the interaction of two opposing forces, the mechanism of faithful gene replication, and the tendency of variability to be introduced from time to time into a gene pool. The interaction of these two forces is mediated by the environment, so that organisms become specifically adapted to a set of conditions—adaptation in that sense is not the result of a haphazard set of events.

The work of Wright, Fisher, and Haldane found an extension from 1940 to 1960 in three directions that brought the synthesis to its final completion. One was toward the more careful genetic analysis of natural populations, combining field studies with laboratory breeding and cytogenetic analyses (cytogenetics is the study of chromosomal structure as it relates to hereditary traits in the phenotype). Another was toward a more thorough understanding of the problem of species, their nature, and their characteristics under field conditions. Still a third was toward the foundation of more complex populational models, including not only genetic and evolutionary but also ecological factors and their myriads of interconnections. We will consider in detail only one of these directions: the genetic analysis of natural populations.

When Theodosius Dobzhansky came to the United States in 1927 to learn the breeding and cytological methods of the Morgan group at Columbia, he brought with him the experience and tradition of the Russian school of population genetics. Morgan and his group were not much interested in population genetics per se, although they saw its implications for evolutionary theory, at least in broad outline. Dobzhansky, however, combined the populational approach of Chetverikov with the breeding and cytological methods of the Morgan School to provide thorough analyses of the genetic composition and changes in composition of natural populations over time. Not a mathematical model builder such as his Russian colleagues, Dobzhansky was nonetheless well versed in statistics as a means of rigorously analyzing field and laboratory data. His interest lay primarily in the natural population and what could be learned about its heredity and evolution, unlike Chetverikov and others of the Russian school, whose interest had been more in the model itself, the real population serving largely as a testing ground for theoretical constructs. Dobzhansky's method involved a combination of field sampling studies and subsequent laboratory experiments on *Drosophila,* including cytogenetical analysis.

Surveying populations of *Drosophila pseudoobscura* taken at regular geographic intervals from California to Texas, Dobzhansky found that different populations could be characterized by different frequencies of certain chromosomal rearrangement patterns. These patterns were detected by cytological study as changes in banding arrangement but had no visible phenotypic effect by which they could be recognized or spotted. More interestingly, he found that the frequency of four such arrangements changed within any one population in a regular, cylic pattern over the course of the year. Dobzhansky accounted for these changes on the basis of natural selection and proceeded to use his observations as the basis for an experimental demonstration. Until the 1940s it had been thought generally that evolution by selection was too slow a process to be observed by human beings in their lifetime—that it was a process whose effects could be seen only in the span of geological time. Dobzhansky's suggestion that seasonal fluctuation in certain genotypes was produced by selection thus almost seemed preposterous. He pointed out, however, that different chromosomal arrangements very likely had differential adaptive capacities, one type being more adaptive to, for example, cold weather, and another to warmer weather. An important quality of this hypothesis was that it could be tested. Dobzhansky designed a simple experiment involving what was called a population cage (a cage of very fine mesh screen) in which a population of *Drosophila* could be raised under a variety of environmental conditions. Starting with a population of several hundred flies heterogeneous for several (or all) of the chromosomal arrangement types, Dobzhansky found that at high temperatures (25°C) after 10 months, the so-called ST chromosomal arrangement increased from 10 percent of the population to between 60 and 70 percent; at low temperatures (around 16°C) there was no change in the frequency of one type over another. He reasoned that one gene arrangement proved to be more adaptive under one set of temperature conditions, while another arrangement proved more adaptive under another. He was witnessing evolution right before his eyes—in the space of a few short months. Temperature was in this case the selective agent, and the genetic variability was the various arrangements of genes known to occur on one of the chromosomes.

Dobzhansky's experiments, as an example of the extension of earlier work in population biology, emphasized two important points. The first was that at last even the theory of evolution by natural selection could be subject to experimental and quantitative tests. As Dobzhansky wrote at the beginning of a paper on his *Drosophila* studies in 1947:

"Controlled experiments can now take the place of speculation as to what natural selection is or is not able to accomplish. Furthermore, we need no longer be satisfied with mere verification of the existence of natural selection. The mechanics of natural selection in concrete cases can be studied. Hence the genesis of adaptation, which is possibly the central problem of biology, now lies within the reach of the experimental method."

To Dobzhansky and many other biologists at the time it seemed that a problem that had been refractory to experimental study for over a century had at last yielded to rigorous analysis. The experimental and quantitative methods so successful everywhere else in biology had at last proved successful in evolutionary theory as well. The final bastion of the descriptive morphologists and naturalists had given way to the new methodolgy.

The second point that Dobzhansky emphasized was that what appeared to be adaptive were not individual genes or their phenotypic traits but, instead, the whole complex of genetic traits carried on a chromosome, a part of a chromosome, or within the entire complex of chromosomes in the population. As he wrote in 1947:

"Natural selection may operate in a variety of ways. The chromosomal types may be characterized by differential mortality, or differential longevity, or fecundity, or differences in sexual activity, or combinations of two or more of these and other variables. The adaptive value of a chromosomal type is the net effect of interaction of all the variables."

Adaptation, the key problem in evolutionary biology from Darwin onward, was the result of a multitude of interacting genes and was not something simply characteristic of individual genes. By studying chromosomal groupings of genes, Dobzhansky was at one and the same time experimental, rigorous, and materialistic, without being mechanistic and reductionist. Adaptation was the result of integrated processes (gene actions) that could only be understood if viewed as part of a whole: the total environment (physical and biological) in which the organism lived.

We can discuss only briefly the other two lines of influence—the species problem and mathematical population biology—that developed out of the evolutionary synthesis of the 1930s. Both these lines illustrate the same pattern of development as Dobzhansky's field and laboratory studies: a growing recognition of the complex interaction involved in populational phenomena, and the origin of new methods and concepts for dealing with these complexities. Attempts both prior and subsequent to Darwin to understand what a species was had been characterized by mechanistic or idealistic thinking. Darwin's nominalist position was essentially mechanistic—that species do not exist, what existed were only individuals interacting with one another. In the twentieth century, however, this view came under increasing attack until, in the 1950s, the whole "species problem" was studied and clarified by workers such as Julian Huxley (1887-   ), Ernst Mayr (1904-   ), and George Gaylord Simpson (1902-   ). Their work established that species are real biological entities with their own characteristics and histories. But their approach, especially as exemplified by the writings of Mayr, has been holistic in method and philosophy. Older writers, in Darwin's day and before, who argued the same point (that species exist) generally sought a single principle by which any two groups could be defined as one or two species. For some that criterion had been morphology, for others reproductive fertility, and for still others kinds of adaptations that the organisms showed, or their ecological role. Mayr and others pointed out that no single criterion was going to suffice for defining what a species was. It was necessary, he argued, to adopt a "biological species" definition that included all of these factors, as well as others, in deciding what constituted, from many individuals, what real species groupings existed. The biological species concept was thus a holistic approach to comprehending the material reality of species and their position in ecological and evolutionary processes.

A similar approach characterized the work of the mathematical population biologists who came to prominence in the 1950s and 1960s. Their work was a direct and logical extension of that of Fisher, Haldane, and Wright. They sought to devise mathematical models that would account for more variables in population structure and dynamics than his predecessors in the 1930s. The essential process, however, was the same: mathematical model building. Their models incorporated not only genetic and evolutionary but also ecological parameters, and the complexity that some of these formulations have obtained indicates the numerous factors they have tried to consider. Such models are, of course, complete abstractions. But their starting place, like Henderson's blood buffer conception, was the material reality of a complex system. Their end point, like the nomogram, is also a device for conceptualizing numerous interactions in a nonreductivist way.

Conclusion

The two great syntheses that we have discussed in this chapter, one centering about embryology, the other about evolution, had as their common denominator the developments in genetics that took place between 1900 and 1930. In a substantive way genetics and its close companion, cytology, provided the reference points by which a convergence in ideas about biological processes could begin to take place. The convergence was only begun by the 1930s and 1940s—reaching its fullest dvelopment only with the rise of molecular biology in the 1950s and 1960s. The beginning provided, however, a foundation for the increasing realization that all biological phenomena are interconnected and could be dealth with in similar terms. Those terms were largely derived from genetics, although applied in a new context: namely, of the organism or population as an entity, a reacting system meaningful only in the context of its environmental surroundings.

Earlier in the century embryologists had focused their attention on cellular and molecular questions, trying to determine the exact influence of one or another chemical process on development. The *Entwicklungsmechanik* tradition was highly mechanistic, seeking single-component answers to questions about biological processes. It was also reductivist in approach, seeking to break down the organism into its component parts and studying those parts in isolation. The new school of embryology that rose with the work of Hans Spemann benefited from more recent knowledge of biochemistry and genetics; it was also faced with the apparent dead-end road that cellular approaches to differentiation (as with Roux or Driesch) had encountered. As a result, the new embryology turned to supra-cellular (tissue) levels for studying the developmental process. Questions posed on the cellular level missed the enormous complexities that occurred on the tissue level and, in fact, could provide no clue as to how those activities (tissue level) could take place. Originally something of a mechanist, Spemann became a holistic materialist, seeing the organism as a set of interacting and constantly changing parts that defied simplistic explanations such as that occasioned by the search for *the* organizer substance.

The evolutionary systhesis of the 1930s followed a similar pattern. The evolutionists who abounded at the turn of the century were often highly mechanistic and simplistic in their approach to the problem of natural selection. One school claimed that selection was the all-powerful agent for producing evolutionary change, another claimed that direct effects of the environment were the major agent, while still others favored macromutations. The emphasis on randomness and chance that Darwin himself had suggested was a major component in the evolutionary thinking of many turn-of-the-century workers, an emphasis wholly compatible with mechanistic thinking of physicists in the 1840s and 1850s (compare, for example, its similarity to the kinetic theory of gases). Yet all these approaches, by 1920, had not yielded a rigorous or in any way quantitative treatment of the evolutionary process. The early mathematic population geneticists attempted to remedy this situation by combining Mendelian genetics with biometrics to treat the evolutionary process in a quantitative and yet interrelated way. Yet even the early mathematical geneticists did not wholly escape the mechanistic bias. Fisher, for example, compared the interaction of Mendelian genes in a gene pool to molecules of gas in a container and, applying the gas laws from classical physics, emphasized the randomness and chance events that could, on a statistical level, yield predictable results. Later mathematical biologists abandoned such analogies, concentating on the interactions of a multifactorial system. Genes were no longer treated as hard, immutable, or atomistic units but as interacting parts of a whole—that whole being both the genome of the individual and the gene pool of the population. The mathematical models for population dynamics developed by the 1950s and 1960s were highly sophisticated and holistic attempts to treat evolution in terms of populations in their environments instead of as groups of atomistic individuals. Like development and physiology, evolutionary biology moved, from the 1930s on, away from explanations based on the interaction of unchangeable atomized units to consideration of the dynamics of constantly changing systems.

But, for studies on the more complex level to continue to expand and be meaningful, more had to be known about precise events at the lower levels—that of cells and molecules. Both developmental and evolutionary studies found themselves dealing with cellular and subcellular events by the late 1950s. The key discipline that effected this more complete synthesis was called "molecular biology" and in its rise figured, centrally, "molecular genetics."

CHAPTER VI

The Chemical Foundation of Life: The Growth of Biochemistry in the Twentieth Century

I N ADDITION TO genetics and evolution, the most revolutionary areas of biology in the twentieth century have been biochemistry and molecular biology. Concerned with the study of biological processes at the chemical (and molecular) level, these two fields represent a culmination of the dream espoused by Jacques Loeb and other mechanists at the turn of the century. It was in the study of the typical reactions making up life processes (biochemistry) and in the structure of the molecules by which these reactions were brought about (molecular biology) that modern biologists have come close to realizing the aspirations of their predecessors who sought to understand vital phenomena in chemical and physical terms. This chapter concerns the growth of biochemistry, and the following chapter concerns molecular biology (particularly molecular genetics).

Although the terms biochemistry and molecular biology are used to denote different sets of problems and different methodologies, they also have numerous areas of overlap. Strictly speaking, biochemists investigate the chemical reactions involved in certain living processes (e.g., respiration or protein metabolism) with a special concern for identification of reactants, products, and other substances such as cofactors (like vitamins) or enzymes (organic catalyst), which might be involved. Their interest goes further, however, than mere input-output relationships in chemical reactions. They are also concerned with possible mechanisms by which certain reactions are brought about. This includes especially the problems of catalysis, kinetics (rates of reactions and the factors influencing them), energy exchanges, the number of intermediate steps involved and, more recently, homeostatic and feedback control. Biochemistry comprises a variety of studies that aim to describe, qualitatively and quantitatively, the multitude of chemical reactions that occur within the living cells.

Molecular biology, on the other hand, is more concerned with the structure—the detailed architecture—of biologically important molecules. Traditionally, molecular biologists have studied the atom-by-atom anatomy of macromolecules (i.e., large molecules) such as carbohydrates, lipids, proteins, and nucleic acids found in living systems. Early in the twentieth century, biochemists studied the function (chemical characteristics), while molecular biologists (after their start in the 1920s) limited themselves to the physical properties and structure of these macromolecules. In time, however, molecular biologists and biochemists found that their studies were so complementary that neither could afford to ignore the work of the other. The investigation of living processes at the subcellular level requires attention to both the details of molecular structure and to the complex set of chemical interactions in which the molecules participate. For example, molecular biologists have gained considerable insight into understanding the architecture of proteins by the evidence obtained from classical biochemistry about how these molecules function in a chemical reaction. Biochemists, in turn, have made great strides in understanding the complex kinetics of many reactions by knowledge of the precise, three-dimensional structure of the molecules involved. As biochemistry and molecular biology have converged successfully on a number of problems in recent biology (from protein structure to the genetic code), the boundary between them has gradually begun to disappear.

The growth of biochemistry follows many of the same patterns that we have observed with other fields of biology in the twentieth century. It grew out of a nineteenth-century background that was plagued with speculation, confused terminology, and inconsistent conceptual schemes. In its early development in this century, biochemistry drew much stimulus from the physical sciences, especially chemistry. It both contributed to and was given enormous impetus by Jacques Loeb's "mechanistic conception of life." Finally, its development largely involved breaking away from a qualitative and morphologically based tradition that saw biochemical reactions as inevitably bound up with living (cellular) structure as a means of solving some of its most persistent problems.

However, unlike the other areas discussed in previous chapters, biochemistry was, from the beginning, a hybrid field in which chemistry and physiology (in particular) played a large role. It was always, even from the start, more closely associated with the physical sciences than heredity, evolution, of embryology had been in their early days. This characteristic modified but did not fundamentally change the pattern of development that biochemistry shared with other fields of biology in the twentieth century.

Like general physiology, biochemistry always had a certain allegiance to the experimental approach; hence, it was seldom claimed that any problem in biochemistry was not amenable to laboratory study.

Although its roots lie in the nineteenth century, biochemistry has been essentially a twentieth-century science. As with all sciences, there has been continuity in both content and methodology in biochemical work between the nineteenth and twentieth centuries. Nonethless, the kinds of questions asked and the approaches to answering them were very different in 1930 than in 1880. To understand the nature of these differences, we will examine the development of ideas in a specific area: the problem of enzyme catalysis and its relation to ideas about cellular respiration.

Cellular Respiration

Today, respiration[1] is defined biochemically as the series of chemical reactions by which a living cell breaks down fuel molecules (principally carbohydrates and fats) to produce energy with the release of certain waste products (carbon dioxide, water, ethyl alcohol, or lactic acid, depending on the type of cell.) Chemically, the process is one of *oxidation*,[2] meaning that electrons are removed from the fuel molecules, resulting in the gradual breakdown of those substances, Today, the details of this process are well known for a variety of cells. The electrons removed from the fuel molecules are passed along a series of "electron acceptor" molecules; at the end of the series, the electrons are passed to a final acceptor molecule. For most types of cells, the final acceptor is oxygen. The oxygen we breathe into our lungs, for example, is used by the body cells as a final electron acceptor in the respiratory process. Electron transport takes place in cell organelles called mitochondria; during this process, energy in the form of high-energy phosphate bonds (in the molecule adenosine triphosphate) is generated for use by the cell's many energy-requiring reactions. The general process of respiration can thus be written as:

Fuel Molecules ⟶ Waste (Fuel Fragments) + Usable Energy

The type of respiration in which oxygen acts as the final electron acceptor is called *aerobic respiration*: respiration that needs "air." Aerobic respiration

[1] The biochemical definition of respiration should not be confused with the more general physiological term referring to inhalation and exhalation of air from lungs (i.e., breathing).
[2] "Oxidation" refers to a chemical process and does not necessarily involve the element oxygen. Molecular oxygen does serve as an "oxidizing agent" (i.e., accepts electrons) in many types of reactions, hence the derivation of the term "oxidation."

occurs in all the cells of higher plants and animals, in the cells of most protozoa, algae, and in many types of bacteria and fungi. In some other types of bacteria, in yeast cells deprived of oxygen, and in the muscle cells of higher animals during strenuous exercise, a second type of respiration, called *anaerobic respiration*, occurs. When oxygen is lacking, yeast cells can still extract a certain amount of energy by continuing to remove electrons from fuel molecules. Without oxygen, however, electron transport would come to a halt unless the cells had an alternate acceptor. In yeasts one of the intermediate products from the earlier steps of fuel breakdown acts as the final electron acceptor. The alternative pathways yield ethyl alcohol as a waste product; in some kinds of bacteria they yield acetic acid, and in others, as well as muscle cells of higher animals during exercise, lactic acid (responsible for the souring of milk or for muscle fatigue, respectively). Especially throughout the late nineteenth and much of the twentieth centuries, the term *fermentation* denoted the process of anaerobic respiration. In the older literature, for example, we often find the terms "alcoholic fermentation" or "lactic acid fermentation" referring to yeast and bacterial anaerobic respiration, respectively.

In summary, the equations for aerobic and two kinds of anaerobic respiration are shown below.

Aerobic:

Fuel Molecule $+$ Oxygen $\longrightarrow$ Water $+$ Carbon Dioxide $+$ Energy

$$C_6H_{12}O_6 + O_2 \longrightarrow 6H_2O + 6CO_2$$

(Glucose)

Anaerobic respiration (fermentation):

$$C_6H_{12}O_6 \longrightarrow 2\,CO_2 + 2\,C_2H_5OH \quad \text{(Yeast)}$$

(Ethyl Alcohol)

$$C_6H_{12}O_6 \longrightarrow 2\,C_3H_6O_3 \quad \text{(Some Bacteria and Muscle Cells)}$$

(Lactic Acid)

Although these equations give an overall summary of the input and output of the reactions, they obscure several important facts. (1) The above reactions occur in many steps, with numerous intermediate products. Such stepwise series of reactions are called *biochemical pathways* (or *metabolic pathways*, in relation to those reactions involved in breakdown and buildup of foodstuffs). (2) In all pathways numerous enzymes are involved, usually a specific one for each step. Without enzymes, the rates of all biochemical reactions would be extremely slow, and life as we know it impossible. (3) The oxidation of foodstuffs by the respiratory pathway is related to the generation of usable energy in the form of special phosphate compounds. All of these points are made more clearly in the flow diagram shown in Figure 6.1.

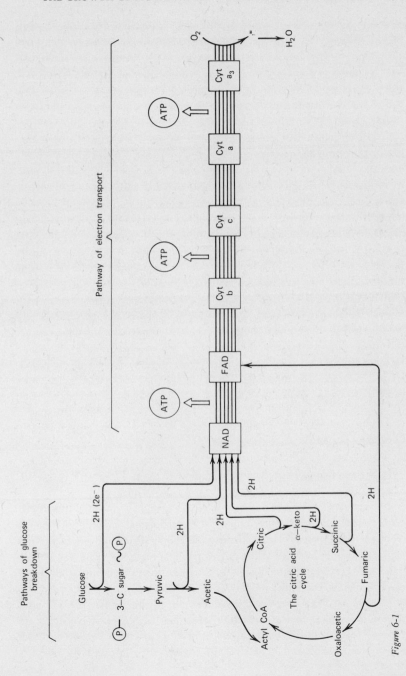

Figure 6-1

Figure 6.1 illustrates a point that must be kept in mind in viewing the history of ideas about cellular respiration. This process, whether aerobic or anaerobic, involves two separate pathways: (1) the breakdown of foodstuffs by the successive removal of paris of electrons (shown on the left, vertically), and (2) the transport of electrons, involving the successive oxidation and reduction (reduction is the *gain* of electrons) of the electron transport molecules (the flavins FAD and FMN, and the cytochromes). The two pathways are ultimately interconnected (as are mos biochemical pathways), so that the rate of one depends on the rate of the other. Those cells, which are obliged to carry out only anaerobic respiration, lack the electron transport molecules. The study of cellular respiration historically has thus involved two quite different lines of investigation: one trying to understand electron transport, another the metabolic breakdown of foodstuffs. Only after 1930 did the two begin to merge in chemical detail.

One of the critical problems in the history of ideas of cellular respiration has been lack of knowledge about proteins in general, and the specific class of proteins called enzymes in particular. Enzymes are organic catalysts that have two very special properties. Like all catalysts, they greatly increase the rate of reaction between two reactants that would normally interact on their own, although much more slowly. Enzymes also have the important property of specificity. Usually, a particular type of enzyme will catalyze a reaction between only one or two, or at most several types of molecules (called substrate). Since investigation of biochemical reaction has required an understanding of the nature of enzyme activity associated with those reactions, the history of biochemistry has been intimately connected to the history of ideas about enzymes, their chemical and physical properties.

The Nineteenth-Century Background

Biochemistry as a distinct discipline developed in the nineteenth century as a part of general physiology. Nineteenth-century physiology, as we have seen, involved the increasing use of chemistry and physics, especially under the influence of the Berlin medical materialists such as Helmholtz and his school. Physiology had also come under the influence of the cell theory of Schleiden, Schwann, and Virchow, which emphasized that the seat of physiological function was individual cells. Although few physiologists specifically sought to study chemical reactions within cells, physiologists became aware of the fact that any doctrines they developed concerning the function of whole organs (or systems) had to be compatible with the function of the units that composed those organs (i.e., cells). It was not only through the conceptual growth of physiology that biochemistry got its start, however. That conceptual growth that characterized physiology during the second half of the nineteenth century reflected itself in the professional organization of physiology as a discipline. The changing institutional position of physiology ultimately provided the atmosphere in which a new, hybrid discipline, biochemistry, could get a start.

By the 1860s physiologists had begun to establish their own identity within European universities. They had not only broken away from subservience to medical anatomy but, in many cases, had assumed leadership over anatomists in medical faculties and research institutes. In fact, physiology as a branch of study came to include not only anatomy but also chemistry and histology. For example, chemistry was being applied to biological problems in three kinds of laboratories within European universities at midcentury. On the one hand, with the rise of organic chemistry, biological problems were being studied within chemistry groups, usually attached to the philosophical (in today's terminology Arts and Sciences) faculties. Medical chemistry offered another area in which chemical aspects of life were studied (in medical schools) and agricultural chemistry, of particular importance in the middle and later part of the nineteenth century, was being investigated within technical schools. This fragmentation was a distinct disadvantage to growth and development of biochemistry as a separate field. The chemical aspects of physiology were themselves investigated under different headings—most notably biological and physiological chemistry which, despite the difference in name, were very similar.

A major barrier to the development of a separate field of biochemistry during the nineteenth century was, in fact, the attitude of the physiologists, who at that time dominated the universities, medical schools, and research institutes. Most physiologists, consciously or unconsciously, wanted to keep all of chemistry, as related to living systems, within the purview of physiology. They wanted to control the kind of research that was carried out and the kinds of questions that were asked about the chemistry of living systems. And when it finally became apparent that a whole new range of chemical questions (within the cell) required investigation, it was physiologists who in fact pushed the formation of separate biochemistry departments. Michael Foster (1836-1907) at Cambridge and Charles Sherrington (1861-1952) at Liverpool were both instrumental in the early 1900s and 1920s, respectively, in establishing biochemistry units in English universities. Physiology was at first the reluctant, but later the enthusiastic, godparent of biochemistry.

Two distinct periods can be discerned in the development of biochemical concepts of enzymes and respiration during the twentieth century. The first, from 1890 to 1925, involved attempts to understand the nature of "ferments" (the nineteenth-century term for what later came to be called enzymes) and their role in cellular respiration. The second period, from 1925 to 1960, saw the development of two lines of work: the study of the structure of specific protein molecules with the ultimate elucidation of the three-dimensional anatomy of a number of proteins; independently, biochemists were investigating the details of intermediary metabolism, the process by which fuel molecules such as sugars are broken down in a stewise manner. From the late 1940s through 1960s the various lines of work developing from 1900 onward were synthesized into a comprehensive scheme of cell metabolism that united biochemistry, molecular biology, and genetics.

To see more clearly how the new field of biochemistry began its development around the turn of the century, it will be useful to look at an example of the older style of physiological chemistry practiced in the latter half of the nineteenth century. An example of the older style is represented by the work of Justus von Liebig (1803-1873). Liebig was a towering figure in nineteenth-century physiology. Stimuluated by the increasing interest in agricultural problems during the early and midnineteenth century in Germany, he applied his training in chemistry to animal and plant physiology. Considered by many to be the founder of agricultural chemistry, Liebig tried to study the physiological problems associated with nutrition. He worked with Friedrich Wohler (1800-1882) on the chemistry of uric acid (a product of nitrogen metabolism) and developed, among other things, the concept that plants are the only organisms that can transform inorganic into organic material. Although this idea was later shown by Claude Bernard and others to be erroneous (animals synthesize many organic substances from scratch), it was an immensely influential idea on the development of nineteenth-century physiology. Liebig directed an institute of physiological chemistry at the University of Giessen and inaugurated the first regularized teaching laboratories in physiology anywhere in Europe. As an institute director, author of many book and journal articles, and editor of *Annalen der Pharmacie und Chemie*, Liebig exerted a widespread influence on a whole generation of students from Europe and the United States.

Liebig was trained as a chemist and studied under the Frenchman Joseph Louis Gay-Lussac (1778-1850) at the Sorbonne from 1822 to 1824. Here he learned the precise analytical techniques that had distinguished French chemistry from the time of Lavoisier. What Liebig brought to biological chemistry was a concern for the analytical process (i.e., for understanding physiological changes within organisms in terms of input-output balances). Liebig emphasized in his early lectures and especially in his two influential books of 1840 and 1842[3] that all changes of a physiological character could be determined by measuring what an organism took in and what it gave off as waste products. Liebig noted that there was a remarkable similarity in the classes of organic compounds found throughout the animal and plant kingdoms—carbohydrates, fats, and proteins. But he also noted that the types of fats found in cows were different from the types found in pigs or in corn. Living things must have a remarkable power of chemical transformation.

"Whatever views we may entertain regarding the origin of fatty constituents of the body, this much as least is undeniable, that the herbs and roots consumed by the cow contain no butter; . . . that no hogs lard can be found in the potato refuse given to swine; and that the food of geese or fowls contains no goose fat . . . . The masses of fat found in the bodies of these animals are formed in their organism . . . ."

To determine how these transformations occur was the central task of the new fields of animal and agricultural chemistry. By knowing these physiological processes, agriculture could become more efficient, less costly, and more productive. Animal and plant growth depended, Liebig reasoned, on the development of a scientific approach to nutrition, itself dependent on knowledge of animal and plant chemistry.

To determine that exact nature of chemical transformations within organisms, Liebig applied a procedure developed by various physiologists, largely French, in the eighteenth century. This procedure involved weighing and measuring all components of an organism's diet and excrement, deducing from the changed proportions of different elements and compounds the nature of the physiological changes that had intervened. Liebig also fed his experimental forms (both animals and plants) controlled diets, in some cases with excesses of certain substances, in other cases with notable deficiencies. He then noted the resulting quantitative and qualitative effects such diets produced on physiological transformations, including careful use of the analytical balance to determine with some precision changes in weights of various substances of the input-output analysis. What he lacked,

[3] *Die organische Chemie in ihre Anwendung aug Agriculutr und Physiologie* (Brunswick, 1840), translated as *Organic Chemistry in its Application to Agriculture and Physiology* (London 1840); and *Die Thierchemie oder die organische Chemie in ihrer Anwendung auf Physiologie und Pathologie* (Brunswick, 1842), translated as *Animal Chemistry, or Organic Chemistry in its Application to Physiology and Pathology* (London, 1842).

of course, were techniques for investigating the actual chemical reactions themselves going on within the organism. He could only infer from his analyses the *possible*, not the actual, changes taking place. For example, Liebig knew that choleic acid was a main constituent of bile salts. Where did this substance come from? Its empirical formula was C_{33} NH_{33} O_{11} (now called cholic acid, a steroid); Liebig reasoned that the molecule could be produced by interaction of hippuric acid, starch, and oxygen in the following set of reactions.

2 Hippuric acids	$2 (C^{18}NH^8O^5)$	$=$	$C^{36}N^2H^{16}O^{10}$
5 Starch	$5 (C^{12}H^{10}O^{10})$	$=$	C^{60} $H^{50}O^{50}$
2 Oxygen	$2 (O)$	$=$	O^2
	Total	$=$	$C^{96}N^2H^{66}O^{62}$

These interace and become transformed into the following products:

2 Choleic acid	$2 (C^{38}NH^{33}O^{11})$	$=$	$C^{76}N^2H^{66}O^{22}$
20 Carbonic acid	$20 (CO^2)$	$=$	C^{20} O^{40}
	Total	$=$	$C^{96}N^2H^{66}O^{62}$

Note that the total input exactly balances the total output. Liebig's chemical training led him to try and account for every atom in such transformations. To do this, however, Liebig invented transformations that could yield balanced equations. The above suggestion that choleic acid was found in the organism from hipperic acid, starch, and oxygen was completely *hypothetical*. There was no experimental evidence to suggest that such a set of combinations actually occurred in any living organism. Despite his emphasis on quantitative analytical chemistry, Liebig's animal and plant physiology was largely speculative. He introduced the procedure of careful measurement on starting and end products on the one hand and of physiological speculation about intermediary transformations on the other. In this way Liebig's influence was in many ways not unlike that of Weismann or Haeckel on their respective fields. Speculation about processes from slim evidence became a hallmark of chemical tradition in physiology influenced by Liebig.

Another aspect of Liebig's influence that was of great importance was his association with the idealistic philosophy of vitalism. In the preface to his *Thierchemie* of 1842, Liebig claimed that we can never know what "life" is, although we can investigate its "vital properties." There is an agency operating in living systems that has no counterpart in the nonliving world.

"Natural science has fixed limits which cannot be passed; and it must always be borne in mind that, with all our discoveries, we shall never know what light, electricity, and magnetism are in their essence, because, even of those things which are material, the human intellect has only conceptions. We can ascertain, however, the laws which regulate their motion and rest, because these are manifested in phenomena. In like manner, the laws of vitality, and of all that disturbs, promotes, or alters it, may certainly be discovered, although we shall never learn what life is."

But Liebig's "vitalism" was double-edged. The preface was far more vitalistic sounding than the rest of the book. Liebig was no *Naturphilosophen*, but he was a cautious materialist who emphasized for his reader a skepticism of ever knowing the final truth about "life." In the last analysis Liebig appeared to say not only "Ignoramus" ("We do not know") but also "Ignorabimus" ("We shall not know"). Such a position seemed to imply that life involved something beyond the physical, further implying it may be supposed, that speculation about internal processes was valid—indeed, the only road. Liebig became a figure, like Weismann and Haeckel, against whom later, less skeptical biochemists could revolt. And revolt they did.

Zymase and Early Theories of Enzyme Action

In 1897 the German chemist Eduard Buchner (1860-1917) discovered a substance called zymase in the extract of crushed yeast cells. Buchner correctly characterized zymase as a "ferment" and showed that it was capable of promoting fermentation of sugar in a cell-free system. It was this discovery, as much as anything else, that can be said to have initiated the development of biochemistry as a discipline separate from general physiology. Buchner and other workers such as Franz Hofmeister (1850-1922) believed that enzymes were proteins and that every chemical reaction within cells was mediated by one or another specific enzyme. In the wake of Buchner's theory, a wave of optimism swept through the biological community as numerous workers sought to interpret all life processes as the direct function of intracellular enzyme action.

Prior to Buchner's discovery of zymase, most biologists, physiologists, and cytologists believed that the complex and rapid chemical processes carried out by cells depended on the intricately organized structure of the cells' cytoplasm. Cytology and histology were thus considered essential adjuncts to understanding the nature of biochemical processes. The significance of Buchner's work lay in the evidence it provided to contradict this view. With the discovery of what was thought to be a generally applicable theory of enzymes, it was no longer necessary to interpret complex processes such as cellular respiration in terms of visible cell structure. Henceforth, cell function could be discussed in chemical terms and studied by chemical methods. Just as experimental embryologists had been able to make great progress in their field by liberating the study of development from a slavish addiction to morphology, so Buchner and a host of younger biochemists, in the late 1890s and early 1900s, sought to understand physiological processes as having a chemistry independent, to some extent, from cell structure. While they did not maintain that cell structure was unimportant, Buchner and Hofmeister saw the cell cytoplasm, with all its attendant structural features (organelles and membranes) as essentially a passive agent that perhaps did no more than compartmentalize the various chemical reactions occurring simultaneously within cells. The reactions themselves were carried out, Buchner claimed, like any chemical reaction; using enzyme extracts they could even be repeated in the test tube.

Although it attracted a considerable amount of attention, the theory of intracellular enzymes met with a variety of objections from the beginning. Chief among these was the assumption that substances extracted from cells reacted in the same way chemically in a test tube (*in vitro*) as they did in the living cells (*in vivo*). This assumption was particularly called into question when a number of biochemists observed as early as 1907 that the overall respiratory rate of tissue *extracts* (in this case muscle cells) was considerably less than that of the intact muscle. By 1910 the same had been pointed out for yeast: while whole yeast cells carried out the respiratory reactions very rapidly, immediately after extraction the rate fell off after a few minutes to only 2 or 3 percent of the *in vivo* values. These findings strongly suggested that something more than simply molecules in solution was responsible for cellular respiration.

Another problem was that Buchner's enzyme theory was based on the idea that all enzymes were proteins, thus assigning to protein a key role in all biochemical reactions. From today's standpoint this may seem like an advanced concept but, in 1900, biologists and chemists alike were very uncertain about the nature of proteins. For example, it was unclear whether proteins were amorphous molecules with no regular structure or highly structured molecules with definite compositions like other organic compounds. While Buchner and his contemporaries knew that zymase was a protein, it was an article of faith to assume (1) that zymase had definite molecular structure that determined something about its role in respiration, or (2) that all biochemical reactions involved proteins such as zymase. Many biochemists around the turn of the century rejected the enzyme theory because they knew so little about the proteins that Buchner assumed a key role. The development of protein chemistry in its relationship to the history of biochemistry will be discussed in more detail in Section 4. Suffice it to point out here that lack of definite knowledge about proteins was a major factor in the opposition that mounted against Buchner's enzyme theory.

The criticism leveled against the enzyme theory by the early 1900s led to a shifting mood among biochemists. This mood brought the revival of ideas associating respiratory activity with cell structure and a growing suspicion of attempts to portray such complex life processes solely in terms of *in vitro* chemical reactions. By 1912 the physiologist Max Rubner (1854-1932) could claim that enzyme action was restricted to a very few simple types of chemical reactions in cells, while the more complex and fundamental processes such as respiration represented "vital" activity. To workers such as Hans Driesch, "vital" meant something metaphysical, beyond the laws of physics and chemistry. To others, however, it meant only the material, structural organization associated with living systems.

It was about this time, between 1908 and 1910, that Otto Warburg (1883-1970) began to investigate the problem of cell respiration. A brief investigation of his entry into the field and the course of his intellectual development reveals a great deal about this transitional period in the early history of biochemistry.

Otto Warburg was the son of a leading German physicist. He began as a student of Emil Fischer, working on the synthesis of polypetides. But Warburg was from the start also interested in biological problems. To gain a biological background, Warburg, like most biochemists at the time, went to medical school, obtaining an M.D. degree at Heidelberg in 1911. During his medical training, Warburg was intrigued by the problem of cancer. In particular, Warburg was intrigued by the observation that cancer cells had a much greater rate of respiration and divided much more rapidly than noncancerous cells. Measuring the rate of respiration accurately thus provided not only a diagnostic tool for determining if cells were malignant but also offered a clue as to the possible chemical events underlying cancer. But it was difficult to study the individual cells of higher organisms, since those cells are bound together as tissues. In searching for a more appropriate cell type in which to study the chemical events of respiration and cell division, Warburg was influenced by the embryological studies of the *Ent-wicklungsmechanik* school—particularly the work on sea urchin eggs being carried out at the Zoological Station at Naples. So, in 1908, Warburg went to Naples to work with Curt Herbst (1866-1946), the long-time friend and collaborator of Hans Driesch and T. H. Morgan. Here, Warburg embarked on a study of the chemical events involved principally in cell division. It was in this work that he came to use measurements of respiratory rate in cells as an indicator of changes in chemical processes.

Several influences affected Warburg between 1908 and 1910. One was the Naples Station, with its kaleidoscope of foreign visitors, and its exciting experimental atmosphere. More specifically, at the Naples Station Warburg came into close contact with the cytological tradition in which Herbst, Driesch, and many others were immersed as part of their embryological work. Through this contact, Warburg picked up a strong sense of the role of cell structure in maintaining and affecting biological phenomena such as development.

In the same years, possibly also through contacts at the Naples station, Warburg was influenced by the school of botany headed by Julius Sachs. Sachs' school had for some decades been studying respiration in whole plant cells, with an eye for the development of quantitative and rigorous experimental methods of measuring and characterizing the respiratory process. And last, Warburg was also greatly influenced by Jacques Loeb, with whom he maintained a strong and lasting friendship. (Loeb, it will be recalled, was also greatly influenced by the Sachs school of plant physiology.) Loeb's attempts to study the physical-chemical properties of sea urchin eggs during artificial parthenogenesis was a model of the kind of work that Warburg wanted to undertake. Warburg and Loeb exchanged many letters and, although their plans to work together both at Naples and at Woods Hole never materialized, they tried in their individual work to carry out similar types of studies. For Loeb, Warburg was an admirable disciple, and for Warburg, Loeb was the acknowledged master of a new fusion between biology and chemistry.

Warburg began studying the respiratory rate of dividing sea urchin cells. He made two observations that tended to discredit the view (held by Loeb, among others) that the cell nucleus was the site of respiration. First, he observed that the respiration of sea urchin embryos at the 8 or 32 cell stage was about the same as that of the uncleaved egg, even though the number of cells (and hence nuclei) was considerably greater. Second, he observed that fertilized eggs that were prevented from dividing also developed the same high rate of respiration as those that were allowed to divide normally. Warburg concluded that the dividing nucleus was thus not the site of respiration. This was 1910. In April 1910, while at the Naples station, Warburg made several unexpected observations that provided a clue as to where respiration might occur. Almost immediately after penetration of a sea urchin egg by a sperm, a "fertilization membrane" can be seen to lift up from the surface of the egg, providing a barrier to the entrance of further sperm. Simultaneously, the fertilized egg shows a rapid increase in oxygen uptake (oxygen uptake is a measure or respiratory rate). A second observation was that an alkaline solution increased the rate of oxygen uptake without apparently increasing the alkalinity of the cell interior (cytoplasm). This suggested that the site of action of the alkaline substance might be the cell surface. Confirming this hypothesis was the observation that organic solvents, known to break down cell membranes, decreased the rate of respiration when added to functional sea urchin cells. Warburg was unable to to state exactly how the membrane itself might act as a respiratory structure, although he assumed it might involve some kind of electrochemical process that produced a selective solubility of the membrane for hydrogen ions. Conspicuously absent from Warburg's theory was any mention of enzymes such as zymase.

Warburg's theory of respiration is a good indication of both the old and the new attitudes prevalent in the emergence of biochemistry in the first decade of the twentieth century. On the one hand, his adherence to a membrane theory of respiration reflects the strong influence that the cytological tradition, with its morphological bias, had exerted on him. At the same time, he viewed the newer zymase theory with suspicion. The nature of "ferments" as a general class was not known, nor was anything clearly established about the role of ferments in respiration. Warburg thus viewed the whole enzyme theory as a nonexplanation. To him it was much like the nineteenth-century theories that "explained" the growth of seeds as caused by a "dormative virtue." Although it may seem ironic from today's perspective, Warburg was committed enough to the mechanistic and physicochemical viewpoint in 1910 that he could not accept what he viewed as the speculative qualities of the enzyme theory as formulated by Buchner.

In 1912 Warburg was making some measurements on yeast cells degraded by reagents such as toluene which, among other things, dissolve membranes. He was surpised to find that the degraded mixture could still respire, although much more slowly than before. Moreover, when narcotics were added to the extracts, respiration was greatly retarded. These findings suggested that the process of respiration might not be exclusively (if at all) associated with membranes, but perhaps occurred independently in the liquid cytoplasm of the cell.

To meet these various difficulties, Warburg developed a complex theory of his own involving what he called the "*Atmungsferment*." Warburg admitted that a large number of chemical reactions within cells were probably catalyzed by enzymes, but he maintained that these reactions were relatively unimportant in the overall life of the cell. In the *Atmungsferment* theory, Warburg combined his belief in the chemical and molecular nature of biochemical reactions with the morphological bias that he had inherited from traditional cytology. In 1912 Warburg saw the *Atmungsferment* as a type of protein that was able to carry out fermentation *in vitro*—although at greatly reduced rates. In the intact living cell, however, the *Atmunsferment* was absorbed onto cellular structures and "organized" so that it reacted at maximum efficiency. Although he did not know what the exact nature of the *Atmungsferment*-cell structure complex was, Warburg was convinced that such a structural organization was necessary for normal functioning of cells. Warburg made an explicit attempt with the *Atmungsferment* theory to combine the purely chemical and purely structural theories of respiratory and enzyme activity. There was no dichotomy in his mind between the two approaches; they were both sides of the same coin. Living processes could not be described by chemistry without cell structure, nor by cell structure without chemistry. Warburg stated his unitary view clearly in a letter to Jacques Loeb in 1914.

"The question always comes to this: cell action or ferment action? Structure action or ferment action? I hope I have demonstrated . . . that there is no dichotomy here at all: both ferment chemists and biologists are right. The acceleration of energy-producing reactions in cells is a ferment action *and* a structure action; it is not that ferments *and* structure accelerate, but that *structure accelerates ferment action.*"

In the next 20 years, between 1914 and 1935, several factors combined to alter Warburg's view of the enzyme theory in general and the process of respiration in particular. One was the development of protein chemistry. Another was the growing recognition of the role of heavy metals in biochemical catalysis, particularly the discovery that *Atmungsferment* itself appeared to contain catalytically active iron. To see how the biochemical concept of respiration developed from Warburg's older, more morphological views, a brief digression is necessary into the growth of protein chemistry between 1900 and 1930.

Proteins and Enzymes, 1900-1930

Proteinlike (albuminoid) substances, such as egg white, mucous, and gelatin, had been recognized since antiquity. In 1838 the name protein (from the Latin *proteus*, first substance) was conferred on albuminoid materials by the physiologist Gerardus Johannes Mulder (1802-1880). By the time Justus von Liebig wrote his *Animal Chemistry* in 1842, it was recognized that proteins were among the most important substances to be found in living systems. The rise of structural[4] (particularly organic) chemistry from 1865 had provided mounting evidence that molecules of all sorts not only had definite composition but also had definite geometrical structures. By the 1870s, it was realized that geometric structure affects the chemical properties of molecules. At the same time, organic chemists were coming to realize that the clearest way to prove the actual structure of a molecule was to synthetize it directly in the laboratory—a realization that gave rise in the 1860s and 1870s to the field known as synthetic organic chemistry.

Study of the proteins was profoundly affected by these various movements within organic chemistry itself. Much information about the physical and chemical properties of proteins was collected between 1860 and 1900. By the turn of the century, for example, the following was known: (1) proteins must be relatively large molecules; (2) proteins acted as colloids in solution; (3) the chemical and physical properties (such as solubility) of proteins could be irreversibly altered by heat, physical disturbance (such as whipping egg white), or by strong acid or alkalai treatment; (4) proteins were intimately involved in nitrogen and urea metabolisms in a large number of animals; (5) proteins existed in a multitude of different chemical types not only within the different kinds of organisms but within the same organism; and (6) proteins could be decomposed into smaller fragments by application of strong acids or bases, by oxidation with certain salts, and by the application of protein-digesting enzymes such as trypsin or pepsin.

In addition, however, some other observations had thrown into confusion any attempt to develop a theory of protein structure. For example, it was found that when proteins were broken down by the application of acids or certain enzymes (the breakdown process was called hydrolysis), a number of fragments were produced of different sizes and compositions. Moreover, these fragments were not always the same from one treatment to the next, even when the same protein was involved. Organic chemists also knew of the existence of small, chemically definable units called amino acids that resulted from the breakdown of protein. The relationship between these amino acids and the large protein, however, remained very much a mystery.

---

[4] Structural chemistry is the term for the branch of chemistry that tries to determine the structure of compounds. Friedrich August Kekulé, who first postulated the ring structure of benzene, was pioneer in the structural chemistry of organic compounds.

By the turn of the century the central problem in protein chemistry could be stated as follows: are proteins substances with a definable, precise chemical structure, or are they aggregates of smaller molecules clumped together in a variety of irregular ways? Related to this question was the ambiguous problem of what chemical role proteins played in biological systems. Two major schools of thought had arisen to "explain" the nature of proteins. One was the colloid school, espoused particularly by Wilhem Ostwald and other physical chemists. The colloid school maintained that proteins were large molecules of indefinite proportion, composed of aggregates of smaller groupings. Since proteins had no definite compositon, Ostwald argued that it was impossible to study their chemical structure. In the past, Ostwald had opposed the growth of structural chemistry and had argued strongly that atoms have no material existence. For this and other reaons the colloid school deemphasized the study of protein *structure*, focusing their research mainly on the physical and chemical properties of proteins in solution.

A second school of thought was that proteins were, like other molecules, composed of atoms arranged in definite proportions. Particularly between 1900 and 1910 this school was dominated particularly by the work of protein chemist Emil Fischer (1852-1919).

After an abortive foray into the business world, Fischer, the son of a wealthy merchant, entered the University of Bonn in 1871 to study chemistry under Frederick August Kekule, master of structural organic chemistry. After receiving his doctoral degree from the University of Strasbourg in 1874, Fischer taught at the Universities of Erlangen and Wurzburg, eventually becoming Professor of Chemistry at the University of Berlin in 1892. For his early work on determining the chemical structures of both the proteins and carbohydrates, Fischer received the 1902 Nobel Prize in Chemistry. Several years prior to that (1899) he had turned to a detailed study of the proteins. Working from the viewpoint of the structural chemist, Fischer was convinced that proteins, like all other molecules, were not indefinite structures but had precise compositon and geometric arrangement that could be determined if the right method were employed.

Fischer became interested in proteins largely through the influence of physiology. As he wrote in a paper of 1906:

"Since the proteins participate in one way or another in all chemical processes in a living organism, one may expect highly significant information for biological chemistry from the elucidation of their structure and their transformations. It is therefore no surpirse that the study of these substances from which chemists have largely withdrawn . . . has been cultivated by the physiologists in ever-increasing degree and with unmistakable success.

Nevertheless there was never any doubt that organic chemistry, whose cradle stood at the proteins, would eventually return to them. There has been, and still continues to be difference of opinion only regarding the data when the collaboration of biology and chemistry will be successful."

It was in the application of structural chemistry to the study of proteins that Fischer saw the possible future collaboration of biology and chemistry. He directed much of his efforts in the next 15 years to this goal.

When Fischer began his work on proteins, about a dozen amino acids were known to result from the breakdown of proteins. Fischer was convinced that amino acids were the building blocks of protein molecules. The basic problem that he faced was to deduce the structure of the intact protein from an analysis of the cleavage products. In fact, that has been the central task of all protein chemistry up to the present time. The major difficulty in carrying out this task was that no satisfactory methods existed for separating the various cleavage products from one another. In 1901 Fischer introduced an extremely important technique. He found that chemical modification of the cleavage products (i.e., the individual amino acids) by a process called esterification made it possible to separate one amino acid type from another by distillation without altering their overall chemical compositions. By using this method, Fischer not only showed that a number of amino acids existed as breakdown parts of proteins, but also by getting relatively pure samples, he could estimate the quantity of each type of amino acids in a given protein. Fischer then combined this technique with a series of laboriously developed procedures whereby he could synthetize individual amino acids and join them together into small units called dipeptides, tripeptides, and so forth. From these studies, Fischer developed the theory of the "peptide bond," the chemical linkage by which all amino acids are joined together to form their respective proteins.

By 1907 Fischer had been able to utilize his synthetic methods to produce an 18-amino acid unit (composed of 3 lucine and 15 glycine units) that, because of its size, he called a "polypeptide." Many new chemical methods were developed by Fischer and others in working toward this goal. Fischer was convinced that the polypeptide that he synthetized artifically had many properties similar to natural polypeptides (although natural proteins have many more than 18 amino acids). As he wrote in 1907:

"One cannot avoid the impression that this (18 amino acid unit) represents a product quite closely related to the proteins, and I believe that with the continuation of the synthesis . . . one will come within the protein group."

Fischer combined two very different but complementary chemical procedures in this work: (1) the breakdown of native proteins into individual
amino acids and the separation of those amino acids to yield a quantitative
analysis; and (2) the rejoining of amino acids together to produce proteinlike
polypeptides. It was the classic method of organic chemistry (derived initially from physics) being introduced into biologically related investigations.

There were some problems, however, with the development of protein
theory during the first decade of the twentieth century. One was the fact that
Fischer's methods only allowed him to combine certain amino acids into
polypeptides. His first 18-unit polypeptide chain contained only two kinds
of amino acids (leucine and glycine). Most native proteins, on the other
hand, contain a wide variety of amino acids. The second was that his separat-
ory procedure, while allowing quantitative analysis of the number and kind
of amino acids of native proteins, yielded no information about the physical
configuration that these units held in the intact chain.

The finding that any particular kind of protein always yielded roughly the
same amount of each type of amino acid indicated that proteins had definite
and repeatable structures. As more and more organic chemists utilized
Fischer's procedures, it became clear that the colloid school had little to
stand on in claiming the indefinite aggregate nature of all proteins. Furthermore, Fischer's unalterable faith in the structural reality of atoms offered a
more certain and enticing pathway for research than the agnostic and
cumbersome idealism of the physical chemists. Jacques Loeb, among others,
found Fischer's work stimulating and impressive. It showed, according to
Loeb, that living processes have the possibility of being reduced to purely
chemical interactions.

Fischer's molecular view of proteins became an important ingredient to
the rise of enzyme theory in the first two decades of the twentieth century.
Many "ferments" were recognized to be protein or proteinlike substances.
Yet, without any clear indication that proteins were in fact molecules of a
definite composition, it had always been diffficult to interpret their catalytic
role in strict chemical terms. Those more progressive and enthusiastic young
chemists who had espoused Buchner's zymase theory after 1897 found in
Fischer's work a notable confirmation of the view that all biochemical and
physiological reactions are controlled by specific enzymes.

There is an irony, however, in Fischer's work on polypeptides during this
period. Fischer himself saw little of the immense implication that his theory
of polypeptides could have for an enzyme theory. Although he believed
that proteins in general fit well the polypeptide theory he was evolving, he

did not believe (a) that all, or even most, enzymes were proteins, or (b) that the specificity of enzymes resided in their protein portions. For example, in 1907 Fischer (and others) had argued that enzymes were not really specific proteins but small catalytic molecules that adhered to some sort of proteinlike colloidal substance. In fact, in the same year, Fischer became embroiled in a controversy over whether the enzyme saccharase (which catalyzes the breakdown of certain sugars) was a protein at all. Fischer argued that the protein component of saccharase appeared to be nonspecific and chemically separable from the catalytic component. The protein thus appeared to be sort of passive carrier having nothing to do with the process of catalysis. Although Fischer was not an adherent of the colloid school, he could not, in the early 1900s, however, detach himself completely from some of its ideas. There was a strong intellectual barrier in the early decades of the present century to seeing catalysis as carried out by large molecules such as protein. Somehow, organic catalysts appeared to most workers at the time to be very small molecules, that is, like inorganic catalysts such as platinum.

The legacy that Fischer's work left behind was severalfold. His exact quantitative methods set a new standard both for organic chemistry in general and for protein chemistry in particular. These methods began to have considerable influence within the areas of biochemistry that, in the early twentieth century, had contact with work on proteins. Fischer's methods, especially artificial peptide synthesis, had also opened up a new way to approach the study of proteins as molecules of exact and definite proportions (i.e., analysis and synthesis). At the same time, Fischer's work had not been able to show in any conclusive way that large, naturally occurring proteins were linked together by peptide bonds as was the case with the synthetic 18-unit peptide generated in the laboratory. In the two decades following 1907, it was still quite possible to view proteins as indefinitely structured molecules that, in the natural state, were very different from the artificial substances Fischer had generated in the laboratory. Between 1910 and 1935 several lines of work led to an increasing acceptance of the view that proteins were specifically structured molecules of definite atomic composition.

One line centered around studies of the size and shape—the physical properties—of proteins. Fischer and his contemporaries had argued that no protein probably had a molecular weight greater than 5000. But, in 1971, the Danish chemist Soren Sorensen (1868-1939) did a series of osmotic pressure studies in protein solutions. From his calculations Sorensen claimed that some proteins, such as egg-white albumin, had a molecular weight greater than 34,000 (today's value, about 45,000). In 1925, however, the

Swedish physical chemist, Theodor Svedberg (1884-1971) introduced a wholly new technique that revolutionized the determination of molecular weights in large molecules. He developed an apparatus called the ultracentrifuge in which solutions of various molecules could be spun at very high rates of speed. The more dense a molecule, the faster it will settle out of a solution at a given speed of rotation. By making a few assumptions about size and density, Svedberg was able to calculate accurately the molecular weights of proteins. He showed that proteins such as hemoglobin had molecular weights in the range of 66,000 and higher. The method of utltracentrifugation, still used today as one method of obtaining accurate molecular weight values, thus showed that proteins were very large molecules—much larger than had been assumed in the early years of the century. And, since the same values could always be obtained for the same protein, Svedberg's findings were a further indication that protein molecules possessed definite size and composition.

By the mid-1920s the colloidal school was beginning to die out. Still, a number of problems confronted those who sought to demonstrate more positively the definite composition of protein molecules. One was the difficulty in purifying proteins—of making sure that any sample being analyzed was only a single type of protein molecule. Another was the problem of adequately separating breakdown products (separated amino acids) of large proteins. A third was the impossibility of synthesizing peptides that contained all the amino acids in the laboratory. Between 1920 and 1950 a series of developments began to solve some of these problems.

One of the developments was the perfection of a method for crystallizing proteins. The first successful crystallization was brought about by James B. Sumner (1887-1955) who, in 1926, produced crystals of the enzyme urease (which converts urea to carbon dioxide and ammonia). In 1935 the American chemist J. H. Northrup (1891-   ) successfully crystallized another protein, the common enzyme pepsin (a protein-digesting enzyme). Crystals are latticework arrangements of molecules that can form only if a single type (single size and shape) of molecule is present. The ability to crystallize a substance means that it is possible to obtain that substance in virtually pure form.

Another development was the perfection of methods of peptide synthesis by the German chemist Max Bergmann (1886-1944), and Leonidas Zervas (1902-   ). In 1932 they developed techniques whereby a variety of amino acid types could be successfully incorporated, at will, into artificial peptides. The importance of their work lay in the fact that by applying these methods they could synthesize polypeptides that more closely resembled in size and composition of naturally occurring proteins than had ever been possible before. The ability to synthesize a particular substance has always been a powerful tool in chemistry; if it is possible to synthesize a given molecule by laboratory techniques, powerful support is provided for the theory of that molecule's atomic arrangement.

A third development involved the perfection of the technique of chromatography. Discovered in 1906 by the Russian chemist Michael Semenovich Tswett (1872-1919), chromatography is a means of separating a mixture of substance by taking advantage of their (relatively) differing capacities to pass from a solution and become adsorbed onto a solid surface. For example, a solution of different dyes is allowed to move up a strip of absorbent paper or poured through a column of packed resin or starch. Each type of dye, because it is molecularly different from each other type, will have its own characteristic tendency for staying in solution (as it moves up the paper or down the column) versus adsorbing out onto the solid substance. Thus, after a period of time, all the molecules of each type of dye will adsorb out at the same place along the paper or column, creating a homogeneous set of bands of (in this case) different color dyes. In the early 1940s, two English protein chemists saw the value of this technique for separating the various breakdown products of protein. Archer J. P. Martin (1910-   ) and Richard L. M. Synge (1914-   ) applied chromatography to the heterogeneous mixture of single amino acids, dipeptides, tripeptides, and so forth, resulting from hydrolysis of natural proteins. This method yielded startlingly accurate and simple results. The main significance of chromatography lay in the ease with which it subsequently made possible the isolation and quantitative determinations of the different types of amino acids contained in any protein.

As a result of these developments, a group of chemists in England, under the directions of Frederick Sanger (1918-   ), undertook in the middle 1940s to work out the complete, detailed arrangement of amino acids in one protein. They chose the hormone insulin, partly because of its ubiquity (it occurs in all mammals) and partly because of its small size relative to other proteins. Sanger and his group used a variety of methods to break down insulin, thus achieving, by each method, different kinds of breakdown products. For example, using the protein-digesting enzyme pepsin, they always got a preponderance of fragments ending in certain amino acids; using trypsin (another protein-digesting enzyme), they got fragments usually ending in another type of amino acid. Using a strong acid, on the other hand, gave a much more random assortment of fragments. They assumed from this that (1) acid hydrolysis breaks peptide bonds randomly, whereas

(2) protein-digesting enzymes are specific, cleaving peptide bonds only around certain amino acids. So far, there was nothing new about Sanger's chromatographic method to separate the breakdown products. Once Sanger and his colleagues had isolated each fragment, they could then chemically characterize it (identify the type or types of amino acid it contained) and determine the quantity present. In this way, by identifying hundreds of fragments and determining the frequency of each fragment, Sanger was ultimately able to state not only the number and kind of each amino acid in insulin, but also the specific order by which they were linked together.

The difficulty of this task in 1945 has been likened by Sanger himself to trying to reconstruct an intact automobile by examining only fragments in a junkyard. The key, according to Sanger, is finding fragments of two or more amino acids joined together—such as finding in the junkyard fragments containing a wheel and an axle attached together. This suggests these two parts are attached in the intact car. Similarly, when three amino acids are found together a number of times among the breakdown products of insulin, Sanger could conclude that these three amino acids exist together in the intact molecule. Without chromatography, it would have been impossible to separate two-amino acid from three-amino acid fragments with any certainty. Thus, by the mid-1940s, Sanger and his group produced a picture of the amino acid sequence of insulin for one species, the cow (such as that shown in Figure 6.2). The molecule consists of a total of 51 amino acids, arranged into two polypeptide chains (labeled *a* and *b*), cross-linked by disulfide (sulfur to sulfur) bonds. Sanger's work for the first time showed that proteins were polymers of amino acids linked together by peptide bonds.

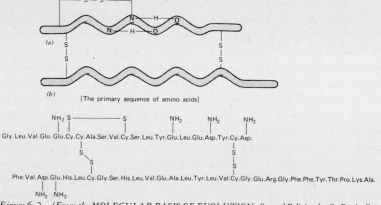

Figure 6-2   (From the MOLECULAR BASIS OF EVOLUTION, Second Edition by C. B. Anfinsen, 1963. By permission of the author.)

Yet Sanger's work established only the amino acid sequence of insulin. It did not provide any clue to the three-diminsional structure of the *a* and *b* chains: were they long and ropelike, coiled, folded back on themselves, and spherically arranged? Development of concepts of the three-dimensional structure of proteins followed two lines during the 1930s and 1940s. One was the theoretical work of Linus Pauling (1901-   ) on protein coiling; the other was the empirical and theoretical analysis of x-ray diffraction data on specific protein cyrstals carried out by Max Perutz (1914-   ) and John C. Kendrew (1917-   ).

In the mid-1930s, American chemist Linus Pauling developed a new view of protein structure out of his general theory of chemical bonding. A theoretical chemist of enormous insight, Pauling showed that a number of types of chemical bonds act within and between molecules and molecular groups in a macromoleculelike protein. In particular, he evolved the notion of "weak interactions," attractive forces between atoms and groups of atoms that are less strong (and thus more easily broken) than so-called covalent bonds (peptide bonds are covalent). Whereas a slight increase in factors such as temperature or acidity of a solution will not be sufficient to break peptide bonds, they were shown to have serious effects on some physical and chemical properties of proteins. Yet, as long as the change is slight, reversing the conditions brought a reversion of the protein to its normal properties. Pauling and his colleague A. E. Mirsky (1900-   ) proposed in the 1930s that the amino acid chain of proteins is not simply a long, stringlike molecule but is folded back on itself in a variety of ways. The folds, they argued, are always precise, being maintained by weak chemical bonds between side groupings of the amino acids. One type of weak interaction, called hydrogen bonds, was particularly prominent. They gave the amino acid chain one of its most important geometrical configurations, the so-called alpha helix. The alpha helix is most generally represented as a coil, much like a spring. Hydrogen bonds are crucial to the folding of the protein into the alpha helix. Pauling and Mirsky's model agreed well with chemical data, such as that obtained from studies of denaturation and renaturation by heat and acidity; it also agreed well with later data from x-ray diffraction studies of crystalline protein. The significance of the alpha-helix concept was two-fold: (1) it provided the first clear evidence that proteins had rather precisely arranged three-dimensional structure, and (2) it provided a method that protein chemists were to find more and more useful for determining molecular architecture: chemical model building. As a result of Pauling's work it was established by the early 1940s that proteins had a three-dimensional structure that was just as important in determining their function as the specific arrangement of amino acids. It had become quite clear that every type of protein had a definite molecular configuration; the colloidal concept retained little influence on chemists or biochemists by this time. Further developments in the field of protein chemistry, principally work on x-ray crystallography, will be discussed in relation to the molecular biology of the period 1940 to 1960 in Chapter VII.

These advancements had a significant impact on the developing concepts of metabolic pathways in the 1920s, especially with regard to the idea of enzyme catalysis. It was partly from his own experimental work and partly from changing ideas about the highly specific nature of proteins that Warburg began to evolve a new conception of the chemical events involved in cellular respiration.

Warburg and Biological Oxidation

Warburg's studies on respiration suggested that the *Atmungsferment*, present in the egg and bound to some cellular structure, was responsible for the cell's ability to rapidly oxidize foodstuffs. In the period subsequent to 1910, however, several additional discoveries began to alter Warburg's views.

The initial idea came from work in his own laboratory. In 1913 Warburg's young assistant, Otto Meyerhof (1883-1951), was using the standard technique to measure how much carbon dioxide was produced by respiring sea urchin eggs: displacement of the CO_2 from solution by tartaric acid. Meyerhof noted that tartaric acid, as well as citric acid, decreased the rate of oxidation, while the tartaric and citric acids were oxidized. It was known at the time that both tartaric and citric acids were able to chelate heavy metals (i.e., bind the metal ions and remove them from the reaction site). Warburg was surprised by these results and concluded that the cell substance contained a catalyst that brought about the oxidization of the otherwise stable tartaric and citric acids. This catalyst seemed to be strikingly similar to the *Atmungsferment*. Further investigations of this remarkable phenomenon revealed that the catalyst that hastened the oxidation of tartaric acid (1) was heat-stable (i.e., continued to work even after being subjected to high temperatures); and (2) appeared to contain the element iron. To test the hypothesis that it was in fact the iron (known to occur in trace amounts in the sea urchin egg) that was acting as a catalyst for oxidation, Warburg and Meyerhof added iron salts to a suspension of sea urchin eggs. They observed an increase in respiratory rate (as measured by amount of CO_2 produced).

Doubling the iron salts doubled the rate of respiration. It seemed clear that iron might be the actual catalyst in some way.

Although Warburg saw that these results could be interpreted as support for the enzyme hypothesis of respiration, he was, at first, more conservative. He could not bring himself to admit the possibility that a purely chemical agent (iron atoms) could catalytically promote a biological process without involving biological structures. Thus the old *Atmungferment* theory continued to loom large in his thinking. Meanwhile, the role of iron in respiration greatly intrigued Warburg and he set out to study it more carefully.

One of the first problems that Warburg undertook was to develop a more accurate method for determining respiratory rate. The older method, involving measurement of amount of carbon dioxide produced by tartaric acid displacement, was clumsy and subject to considerable inaccuracy, especially since CO_2 is soluble in water. In the early 1920s Warburg decided to measure not the carbon dioxide produced by respiration, but the oxygen consumed. To do this, he designed an apparatus known as a respirometer, an adaptation of which is shown in Figure 6.3. In this appartus the cells or free enzyme systems whose respiration is to be measured are placed in a special reaction vessel (shown at the right and in the enlarged insert). In the main body of the reaction vessel is the medium containing living cells or substrate to be oxidized. In the center of this vessel is a well containing a substance (such as potassium hydroxide, KOH), which absorbs carbon dioxide; the reaction vessel is attached to a manometer tube with mercury in the U-shaped arm, the entire apparatus being completely sealed off from the outside. As respiration proceeds, oxygen is used up and carbon dioxide is produced. Since the carbon dioxide is absorbed by the KOH in the center well, a partial vacuum is produced that causes the fluid in the right-hand arm of the manometer tube to rise. The rate at which this rise takes place is proportional to the amount of oxygen consumed and thus indicates the rate of respiration.

A problem that had beset many workers studying cellular respiration in the early decades of the twentieth century was that created by using large chunks of tissue (at the time it was impossible to measure respiration in single cells). The problem was that the innermost cells of a chunk receive nutrient (or anything else added to the medium) only by diffusion; this means that cells of the whole chunk are seldom all respiring at the same rate; it also means that there is no certain way to determine actual starting time for any respiratory reaction. Such limitations are especially notable when the aim is to study specific chemical reactions within the cells themselves.

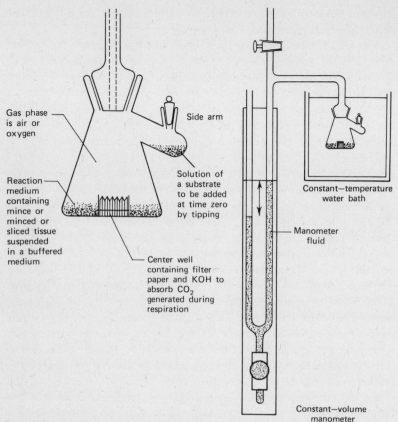

Gas phase is air or oxygen

Side arm

Reaction medium containing mince or minced or sliced tissue suspended in a buffered medium

Solution of a substrate to be added at time zero by tipping

Constant—temperature water bath

Manometer fluid

Center well containing filter paper and KOH to absorb CO_2 generated during respiration

Constant—volume manometer

Figure 6-3   (From Lehninger, BIOCHEMISTRY, Worth Publishers, New York, 1970, page 279. Reprinted by permission.)

Warburg introduced a very simple method of slicing tissues only a few cell layers thick. This procedure minimized the diffusion problem, insuring that within a very short time after a slice was placed in contact with medium, all the cells would have absorbed nutrient or other substances in the reaction vessel. At the same time, the reaction could be stopped almost instantaneously by the addition of an inhibitor. In conjunction with the respirometer, the tissue slice technique made it possible to determine with great accuracy the exact starting or finishing point for a biochemical reaction within living cells.

In determining the role of iron in relation to respiration in general, and the *Atmungsferment* in particular, Warburg combined measurements on living cells with (1) studies of chemical (nonbiological) model systems, and

(2) spectroscopic analyses of the active respiratory pigments isolated from respiring cells.

The use of model systems from physics or chemistry as a means of understanding biological processes has always been a tempting although often misleading approach. Loeb employed it in his analogy of animal behavior to the selenium-eyed dog, and neurophysiologists in the last 30 years have often used electrical circuitry models to describe interconnections in the animal nervous system. The use of such analogies is based on the view that processes that seem similar in that overall behavior may function in the same way in their innermost details. Sometimes this is true. However, many more times it is not. To take analogies literally is an error into which biologists have been prone to fall. But Warburg was lucky in this instance.

For his model system, Warburg used an iron catalyst obtained by heating hemoglobin to incandescence—producing a substance called hemin charcoal.[5] This substance could catalyze a number of oxidation reactions *in vitro*; as such it behaved much like the respiratory enzyme Warburg had isolated in living cells. In fact, the two systems appeared to have much in common. For instance, if hemin charcoal were added to a solution of amino acids in the presence of molecular oxygen, the latter was absorbed and transferred to the amino acids, producing various oxidation products. More amazing, Warburg found that carbon monoxide, and cyanide, long known as respiratory inhibitors, also blocked the oxidative action of the hemin charcoal. Just as with natural respiration, traces of either carbon monoxide or cyanide were sufficient to completely inhibit the hemin charcoal's catalytic activity. Moreover, inhibiting effects of carbon monoxide turned out to be reversible for both systems—that is, the inhibiting effects of carbon monoxide could be reversed in both cases by increasing the amount of oxygen available to the system.

A final similarity lay in the differential effects of light and dark on carbon monoxide inhibition of the two systems. It had been known since the late 1890s that the inhibiting effect of carbon monoxide on iron-containing substances, including hemoglobin, is reversed under illumination. In the dark, carbon monoxide attaches to iron binding sites; in the light, the carbon monoxide is split off from the iron. Warburg subjected both respiring yeast cells and the model system (hemin charcoal) to carbon monoxide and then successively illuminated and darkened the reaction vessel. He found that both systems responded in identical ways, being inhibited and stimulated in the same proportions by dark and light, respectively.

---

[5] Hemoglobin contains iron as part of the heme ring structure; oxygen binds to the iron.

All of these results suggested strongly to Warburg that the iron-containing enzyme he had identified in 1914 in sea urchin, yeast, and other cells was responsible for cellular respiration. Furthermore, he concluded that it was the iron-containing portion of the respiratory enzyme that was active at the catalytic site of the enzyme.

In the middle and late 1920s Warburg was in much the same situation as Morgan and his group had been between 1915 and 1930. The *Drosophila* workers had seen the obvious parallels between the Mendelian theory and the cytological facts of chromosome behavior. But it had not been until 1931, with the discovery of the salivary gland chromosomes, that direct observations had shown that the phenomena were actually part of the same process. Similarly, Warburg needed some direct method for showing that it was, in fact, the iron enzyme of yeast cells that interacts with oxygen, carbon monoxide, and cyanide in the way observed for hemin charcoal. The obvious method would, of course, have been to isolate the iron enzyme from yeast or other cell types in sufficient concentration to perform chemical studies of it *in vitro*. Since hemin charcoal could be prepared in sufficient quantities, such a method would have provided direct evidence for comparison. But it was not a simple matter to isolate a specific protein or enzyme in the 1920s and 1930s. Warburg recognized that the iron enzyme existed in such small quantities that to isolate it would be a very long-range and tedious job. But he did recognize the possibility of using another method: the physical principle of spectrophotometry.

If light is passed through a solution of various atoms, ions, or molecules, each type of substance in solution will absorb certain wavelengths of light and allow others to pass through. A photocell on the other side of the solution can be used to determine the amount of light absorbed at each wavelength. If the amount of light absorbed is plotted on a graph against different wavelengths, an absorption spectrum is obtained. The spectrum is characteristic for each type of molecule in solution. The advantage of this method is that it is independent of the amount of a particular substance in solution, and it can be used with heterogeneous samples. Warburg employed spectrophotometry to study the action of the respiratory enzyme *in vivo*, comparing his results to the *in vitro* data of hemin charcoal. The results of two absorption spectra, one for the combination of carbon monoxide with the respiratory enzyme, the other for carbon monoxide combined with hemin charcoal, are shown in Figure 6.4. As the graphs show, the spectra are remarkably similar. These results provided Warburg with his "crucial experiment." He could now conclude that the hemin charcoal model was not only an analogy; it was chemically very much the

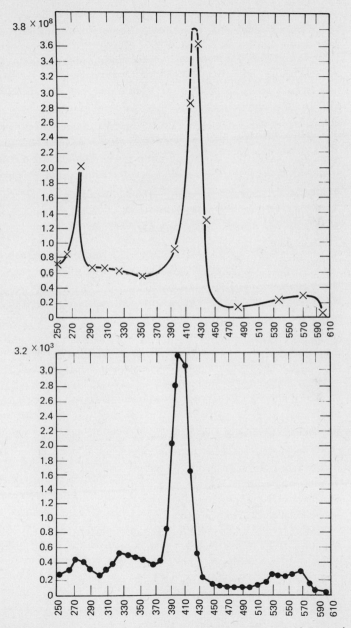

Figure 6-4   (*From Warburg,* Bulletin of the Johns Hopkins Hospital, *1930, Vol. 46. Reprinted by permission of Johns Hopkins University Press.*)

same as the respiratory enzyme. In both systems the heavy metal iron (bound to nitrogen) acted as a catalytic center for oxidation reactions.

"If at this point we look back, we see that in many respects the hemin charcoal has been more than a model. It is true that in the hemin charcoal the porphyrin is destroyed by the heat of incandescence, but still there is the nitrogen of the porphyrin and the linkage of the iron to the nitrogen. Iron linked to nitrogen is the catalyst in both, in the artificial respiration of the hemin charcoal model and the respiration of living cells."

Warburg's work of 1930 was not the first time that a heavy metal had been implicated in either enzymatic processes or in respiration. In 1925, for example, an English biochemist David Keilin had also used spectrophotometric methods to study substances present in insect muscles. As shown in Figure 6.5, Keilin mounted an insect on a microscope stage so that light passed through the muscle leading from the thorax to the wing (the muscle is actually exposed here). An absorption spectrum was produced by the light passing through a prism at the upper end of the microscope. When the insect beat its wings, a group of absorption bands (lines) appeared in the spectrum; the bands disappeared when the insect was still. Keilin concluded that a molecule, or series of molecules, in the cell (which he called cytochrome and which contained iron bound to nitrogen) were responsible for the absorption lines. Keilin thought cytochromes underwent reversible oxidation and reduction, being fully oxidized when the muscle was quiet and fully reduced when the muscle was active and carrying out rapid respiration. It was the reduced forms that produced the dark bands.

Since Warburg knew of Keilin's work, it was natural for him at first to associate the respiratory enzyme with cytochrome. However, he soon found that the two could not be the same. Cytochrome does not combine with either oxygen or carbon monoxide, and it exists in very great quantities in the cell, as compared to the minute amount of respiratory enzyme. Warburg was conviced he had discovered a separate substance; it is what we today call cytochrome a3, or cytochrome oxidase, a distinctly different molecule from cytochromes b, c, and d, which carry out electron transport.

The significance of Warburg's discovery was that it identified a specific enzyme with a complex metabolic process such as respiration, defining the enzyme's role and even the specific part of the molecule responsible for catalysis. Moreover, by using the hemin charcoal model, Warburg had weaned himself and many of his contemporaries from the morphological view of biochemical reactions. Hemin charcoal was a relatively simple molecule that was not bound to any biological (cellular) structures and that

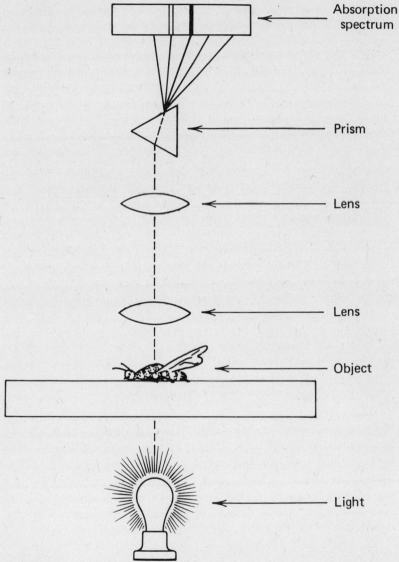

Figure 6-5

yet carried out catalysis. After 1930, the respiratory enzyme was viewed as a distinct molecule whose catalytic function resided in the composition and structure of the molecule itself instead of in some interrelationship between the molecule and biolgical "structure." Catalysis was due, if anything, to molecular instead of to "biological" structure. As a result of Warburg's work, Buchner's enzyme theory—*in vitro* biochemical activity—took on a new life. The phenomena of metabolism, once thought to depend on "vital" organization, could now be seen to exist chemically independent of living, organized cells. Warburg's work had indeed opened up a new vista in biochemical work. For his pioneering work, his identification of the iron enzyme and elucidation of its role in cellular respiration, Warburg received the Nobel Prize in physiology and medicine in 1931.

Warburg's significance for the life sciences in the twentieth century goes deeper than his influence on biochemistry alone. In his experiments, he constantly stressed the importance of quantitative and rigorous work. To Warburg the older style of biochemistry was represented by the general physiology of Liebig, and by the more recent oxidation theory of speculative investigators such as Heinrich Wieland (1877-1967). Wieland, for example, had proposed that biological oxidation of foodstuffs was accomplished by a specific enzyme by means of which hydrogen was removed from the food molecules and brought directly into contact with molecular oxygen to produce hydrogen peroxide. What Warburg objected to about Wieland's theory was not that it turned out to be less than correct in its details, but that it was, from the beginning, a speculative form of biochemistry. Wieland, much like Liebig, postulated an interaction on the basis of knowing what went into and what came out of the reactions involved in oxidation. He hypothesized a likely biochemical pathway, including the hypothetical role for an enzyme thought to be involved, but he had no hard, experimental proof. In criticizing Wieland's work, Warburg wrote:

"The oxidation theory of Wieland . . . [is] premature because when [it] was proposed nothing was known about the chemical constitution of the ferments participating in respiration. [It was a theory] regarding the mechanism of chemical reactions, proposed without knowledge of the participants in the reaction. Such theories cannot be other than erroneous, and they must disappear to the extent that the chemical nature of the reaction partners—in this case the chemical constitution of the ferments—is elucidated."

Like Loeb, Morgan, Sherrington, and others, Warburg saw that if the study of living systems was to progress, it had to be based not on hypothetical possibilities but on measured realities. The hypothetical biochemistry of a Liebig in the nineteenth century or a Wieland in the twentieth was only retrogressive; it obscured details that had to be proven, instead of offering proof itself. For Warburg, too, the reality of the biological system was always paramount. Chemical models for him were only useful to the extent that they corresponded to the measured biological phenomena. Consequently, measurement, careful and detailed studies of the components of biochemical reactions, were always the starting and finishing points of biochemical study. To Warburg, it was essential to know the components of a biochemical pathway before hypothesizing how that pathway might work. Warburg spoke for the new generation of biochemists who broke from both speculative and morphological views of chemical activity in relation to biological organization.

Biochemistry and Philosophy

Warburg's new mode of biochemical work was based on the principles of mechanistic materialism. Revolting from his own attachment to the morphology of the membrane theory, Warburg emphasized the important role that isolated enzymes have in cellular respiration. The enzyme functions to catalyze a given reaction whether or not it is contained within the "living cell." Enzymes became, in Warburg's conception, free-floating molecules carrying out a reaction whenever they encountered appropriate substrate molecules. Subsequent work, especially after Warburg's initial investigation of the respiratory enzyme between 1930 and 1933, showed that the mechanisms of enzyme activity were more complex than Warburg had supposed. Warburg's mature view was mechanistic in that it saw molecules acting more or less separately from the living systems in which they occur. It took the molecules out of context and thus generalized about their activity without viewing the many interconnections that occur in the *in vivo* system. Ironically, Warburg's earlier membrane theory had been less mechanistic in the sense that it emphasized the interaction of chemical molecules with biological structure. The earlier view saw an important role for *organization*, whereas the latter view emphasized the more purely chemical aspect of respiration and catalysis. Nonetheless, Warburg's later views were a step ahead of the earlier. From 1914 to 1915, he could not see as clearly the chemical aspect of respiration because his view was dominated by structural considerations. It was necessary, in some ways, for Warburg to free himself of the notion that biochemical reactions can only take place in conjunction with biological structure. In revolution from this view after 1930, however, he did not come back to its positive points. Warburg's work in this later period can be described as mechanistic in that he did not consider thoroughly the *interaction* between molecules (like enzymes) and larger biological structures (such as membranes of mitochondria). In Warburg's defense, however, it must be pointed out that knowledge of the structure of such organelles as mitochondria and the relationship between that structure and biochemical functioning was only elucidated after 1950 (with the help of the electron microscope).

In the 30 years since Warburg's work, biochemistry has undergone a synthesis similar to that experienced in genetics and evolution in 1930. As Chapter VII will show, the two are intricately connected with the development of molecular biology—in particular, molecular genetics. Between 1930 and 1960, biochemists elucidated the many metabolic pathways involved in the initial phases of respiration: the breakdown of foodstuffs to yield the electrons passed through the cytochrome system to oxygen (see the left-hand, vertical side of the scheme outlined in Figure 6.1). They have shown that virtually every intermediate in this complex pathway is the branchpoint for another pathway, leading to the synthesis or degradation of some other substance produced by cells (e.g., fats are produced by a pathway originating from the intermediate Acetyl CoA in the Citric Acid Cycle, as shown in Figure 6.1). They have identified the specific enzymes catalyzing each individual reaction and have shown that the function of many of these enzymes depends on the presence of small molecules, or enzyme helpers called coenzymes (many vitamins function as coenzymes). They have used radioactive tracers (radioactive atoms inserted chemically into a compound to trace its course through a series of reactions) and agents that block specific enzymes to determine the precise sequence of steps in metabolic pathways. They have shown that genes determine protein structure and, in this way, utlimately control the molecular events in cell metabolism. They have shown that many enzymes are attached to larger structural elements in cells, such as membranes. Biochemists have produced a picture of the cell in enormous detail, but a picture that more and more emphasizes the interconnections between parts, instead of focusing only on the *in vitro* function of isolated reactions. The isolation studies came first out of necessity but, by the 1960s, biochemists were beginning to move away from the view that the characteristics of a reaction *in vitro* were necessarily identical to those *in vivo*.

A clear example of this change in view can be illustrated in the recent work on enzymatic control mechanisms. Biochemists had noted by the 1940s or 1950s that many enzymes appeared to be composed of two or more subunits, each an individual polypeptide chain, usually of one or two identical molecular configurations. The subunits are usually bound together by weak chemical interactions and can be separated by changes in acidity or temperature. The significance of this finding was not recognized until a functional teristic of enzymes *in vivo* had raised considerable curiosity about certain metabolic pathways. In the living cell, most pathways are able to function at different rates of speed—faster when more end-product is needed, slower when the end-product is plentiful. How this regulation takes place was a mystery until it was recognized that the subunits of enzymes play a crucial role. Detailed studies of the structure of these subunits on a few enzymes, starting in the 1950s, showed that one type of subunit was usually structured to interact with the normal substrate of the enzyme; the other type of subunit was specialized to interact with the end-product. When the substrate fit, geometrically, into its specific active site of one subunit, the whole molecule (all subunits) shifted to one geometric configuration (let us say the + form); when the end-product fit into the active site of its subunit, the whole molecule shifted to a second geometric configuration (let us say the —form). In the + configuration, the active site for end-product was distorted, so that end-product molecules could not fit in easily. In the — configuration, the active site for the substrate was distorted so that substrate molecules no longer fit easily into their own active sites. Thus, when there was an abundance of substrate (and less end-product), most molecules of the enzyme were in the + configuration and thus could catalyze the conversion of substrate into end-product. When there was an abundance of end-product, most enzyme molecules were in the — configuration and thus could not catalyze the conversion of substrate to end-product.

This remarkable system of regulation was discovered by studying the *in vivo* pathways in which enzymes take part. The older, more mechanistic view saw enzymes as rigid molecules, preserving their configuration in a constant way and unable to regulate rates of reaction by any mechanism other than the usual one of mass action (i.e., the availability of substrate molecules or the pile-up of end-product). Such a view saw enzymes as analogous to parts of a machine—functioning the same way in all circumstances. The newer view, in which the enzyme's own role in regulation becomes central, is an example of holistic materialism. This view sees the interconnections, the fact that the process controls itself through the interaction of internal, opposed elements (i.e., the opposite effects of substrate and end-product on an enzyme's action); it sees that the course of a biochemical reaction changes the environment in which that reaction is occurring (e.g., by increasing presence of end-product), thereby changing the chemical conditions and the future course of the reaction. It is not that all these elements were missing from the older mechanistic view. But the emphasis was different. The history of biochemistry, as with the other areas of biology discussed in this textbook, illustrates clearly the importance of a mechanistic materialist stage in the development of the modern life sciences: that it is often necessary to isolate individual components (in this case single enzymatic reactions) from a complex interacting system at the outset. The cast of characters, so to speak, must be identified. But the history of biochemistry also illustrates the limitations inherent in the mechanistic approach if applied as an overall philosophy of nature.

Conclusion

The development of biochemistry in the twentieth century, illustrated in this chapter by a detailed study of changing ideas about the enzymatic nature of cellular respiration, grew to a large extent out of the revolt against speculative physiological chemistry of the middle and late nineteenth century. Like other areas of biology surveyed in this volume, biochemistry depended for its development not only on conceptual advances (the overthrow of the colloid school of protein chemistry, for instance) but also on the merger of separate disciplines and the introduction of new technology. In the former category lie the important developments in organic and, later, in protein chemistry; without these orginally quite separate developments, biochemistry would have never advanced beyond the state of simply measuring input and output in living systems (e. g., amount of O_2 consumed and CO_2 released), with no clue about actual chemical pathways or mechanisms of interaction. The development of concrete ideas about protein composition and structure allowed biochemists to understand the importance of specificity and catalysis in the action of "ferments."

The development of new techniques has been an especially important force in twentieth-century biochemistry. Among the relevant techniques have been chromatography, ultracentrifugation, radiocative tracers, Warburg's respirometer, methods of chemical synthesis of peptides, and x-ray crystallography. For example, without Warburg's respirometer, it would have been impossible to determine accurately the starting and stopping times of biochemical reactions *in vivo* and hence difficult to study the metabolic reactions related to respiration. Without methods of chemical synthesis that allow the production of specifically structured molecular groupings, we would know much less about the catalytic nature of enzyme-substrate interactions. Without chromatography and x-ray diffraction (and ultimately the computer), it would have been impossible to build up the complete three-dimensional structure of at least the 10 or 12 proteins fully characterized today. In the history of biology all fields have depended to some degree on advances in technology. But it has always been especially important in the biochemical and physiological sciences.

The unifying principles that evolved under the banner of molecular biology between 1940 and 1965 grew out of a foundation laid by the development of biochemistry in the early decades of the century. That foundation involved a conceptual framework of cells as chemically active units, the reactions inside of which were organized in a stepwise manner, each step controlled by a specific enzyme. The questions that began to rise out of this picture of cell activity provided must of the stimulus for the growth of molecular biology. The questions showed, in a way seldom experienced in the history of biology previously, the fundamental relationships that existed between heredity, physiology, development, and evolution. The history of this synthesis will be subject of Chapter VII.

CHAPTER VII

The Origin and Development
of Molecular Biology

I F THE QUARTER century after World War II becomes known to future generations of scientists by any historical designation, it will be as "the age of molecular biology." While characterized by an enormous growth rate in many areas of natural science, it was in biology that the most far-reaching and revolutionary developments took place. These years saw the flowering and integration of molecular and biochemical studies to a depth and breadth that would have pleased even the most doctrinaire mechanists of the earlier decades of the century. Areas such as embryology, heredity, or evolution, which had been studied previously on a tissue, cellular, or populational level, were gradually shown to have a common foundation in the molecular architecture of specific marcromolecules. Investigations into the structure and function of molecules such as proteins and later nucleic acids opened new vistas for investigating the microstructure of living systems and showed new relationships among broad areas of biology whose common foundations had only been dimly surmised in the past.

As it became clear in the 1940s and 1950s that nucleic acids were the principal substances of heredity and that they seemed to function by guiding the synthesis of protein, the study of heredity once again emerged as a revolutionary and dominant area of twentieth-century biology. The work of the Morgan school had shown that genes could be regarded as material parts of chromosomes, but it had not attempted to study the molecular nature of genes or anything about their biochemical function. The question was real, but the approach to it was crude and obscure. Thus, when appropriate tools and techniques became available for focusing on specific molecular interactions within cells, it is not surprising that genetics once again in the twentieth century offered exciting prospects.

From the start, however, molecular biology was, and is, more than simply molecular genetics. While genetics provided the starting place, the molecular approach spread to other areas such as cell biology, the structure of biological membranes, and the organization of metabolic pathways. Recognizing the many directions that molecular biology has taken, this chapter traces the origins of the new biology principally through one area: genetics. The purpose will be to seek the roots of molecular biology in the early developments, both in content and methodology, which we have already explored in the 1920s and 1930s, and to see how those roots were influenced by factors implicit in, and external to biology itself.

The rapid growth of molecular biology following World War II occurred almost exclusively in three countries: Britain, France, and the United States. This national concentration of efforts arose as much from default of other nations, because of special historical circumstances, as from any positive attributes of those countries most involved. The Soviet Union, for example, was in the grips not only of economic readjustment, but more specifically of the Lysenko movement, which took Russian genetics on an intellectual dead-end for several decades. China, too, was involved in its own internal political and economic struggles, which made rapid scientific and technological development in the Western sense virtually impossible. The war did, of course, have its effects on the Western capitalist countries. In the struggle for neo-colonial and imperial expansion, which in its broadest sense that war represented, the losing coalition (the so-called Axis powers of Germany, Italy, and Japan) was too bankrupt (financially and spiritually) and too depleted of talent (through emigration, death, or imprisonment) to recommence immediately new and ambitious research programs. The winning coalition, dominated in the West by the United States, France and Great Britain, was more ready and able to pick up new strands of scientific developement. All three had inherited the fruits of the intellectural migration from central Europe in the mid- and late 1930s. These émigrés brought with them a rich scientific heritage. The United States in particular had emerged economically and physically more sound from the war than any of the other Allied powers. As the United States strove quickly to inherit the mantle of imperialist domination from the declining European colonial powers, its economy began to grow at an unprecedented rate. Through economic expansionism at home and abroad, financial resources became available for the large-scale support of scientific research that was required for investigations at the subcellular level. Molecular biology is a far more costly scientific enterprise than most of the older areas of biology had been. Molecular biology on a grand scale could only be pursued in affluent economies. The fact that the United States, and Great Britain to a lesser extent, had been centers of classical genetics research for nearly half a century previously gave additional impetus to the continuation of researches in heredity in these two countries.

The Nature of Molecular Biology

In recent years the term "molecular biology" has become fashionable and its uses varied. For some it has become synonymous with biochemistry practiced "without a license," as one writer has remarked. For others it is a shibboleth for getting onto the bandwagon of research funding; for still others it is nothing more than a branch of ultrastructural biology, carried down to the level of molecular architecture. John Kendrew, an x-ray crystallographer, has pointed out that, in fact, although the term is frequently used, many molecular biologists are not even clear on exactly what their subject is about. W. T. Astbury, one of the founders and propagandizers of the term "molecular biology" defined it in 1950 as being

". . . .particularly [concerned] with the forms of biological molecules and with the evolution, exploitation and ramification of these forms in the ascent to higher and higher levels of organization. Molecular biology is predominantly three-dimensional and structural—which does not mean, however, that it is merely a refinement of morphology. It must at the same time inqure into genesis and function."

In the same essay, Astbury added another dimension to his use of the term: "it implies not so much a technique as an approach, an approach from the viewpoint of the so-called basic sciences with the leading idea of searching below large-scale manifestations of classical biology for the corresponding molecular plan." In many ways, however, this definition is incomplete. It only hints at the relation between molecular form and function, the keynote of the present-day conception of molecular biology.

Current "molecular biology" includes not only a structural and functional element, but also an informational element. It is concerned with the structure of biologically important molecules (such as proteins or nucleic acids) in terms of how they function in the metabolism of the cell and how they carry specific biological information. The methods of physics and structural chemistry (such as x-ray diffraction of crystalline molecules, and molecular model building) have been employed to investigate molecular architecture, while biochemistry has been used to determine how large molecules interact with each other and with smaller molecules inside the cell. Historically, three lines of thinking have gone into the formation of molecular biology as we know it today.

1. **Structural:** concerned with the architecture of biological molecules.
2. **Biochemical:** concerned with how biological molecules interact in cell metabolism and heredity.
3. **Informational:** concerned with how information is transferred from one generation of organisms to another and how that information is translated into unique biological molecules.

The structural approach is particularly concerned with three-dimensional problems, the shapes of molecules, and to some extent, how shape determines specific function. The informational approach has been traditionally concerned only with one-dimensional problems, that is, the linear sequence of molecular parts that carry specific biological information. As we saw in the previous chapter, the biochemical approach seems to fall somewhere between these two. In present-day molecular biology all three of these components are fused, so that it is safe to say a complete description of any molecular phenomenon must include structural, biochemical, and informational data. However, in the history of molecular biology in general, and of molecular genetics in particular, this was not always the case. These approaches existed until the late 1950s as three relatively separate areas of study. It was only with their fusion that molecular biology as we know it achieved a preeminent position in twentieth-century biology.

The Chemical and Physical Background to Molecular Genetics

Study of the chemical and physical nature of genetically important macromolecules has had a long history in which methods of both traditional biochemistry and physics have played an important part. Although nucleic acids were discovered by the German chemist Friedrich Miescher (1844-1895) in 1869, their significance for the study of heredity was not fully appreciated until well into the twentieth century. It was the proteins that attracted considerable attention from the early nineteenth century onward. This was partly because of their quantitative abundance (compared, for example, to nucleic acids) in living tissues; but it also came from the emphasis of nineteenth-century thinkers such as T.H. Huxley on the colloidal nature of "protoplasm" as the physical basis of life. Proteins were thought to be largely responsible for determining colloidal properties. In addition, as the structure of proteins became more clear by the 1910s and 1920s, these molecules seemed eminently suitable to carry genetic information. Their component parts (amino acids) could be arranged in a variety of ways, having the potential for enormous complexity of information. The proteins appeared to be the only biological macromolecules having this capability.

With all the flurry of activity surrounding proteins, by comparison nucleic acids received scant attention. Although in the 1860s and 1870s Miescher had pointed clearly to the possible role of nuclelic acids in the hereditary process, his work was largely ignored by several generations of biochemists and geneticists. Like proteins, nucleic acids were recognized by 1930s as polymers. They were composed of four types of nitrogenous bases: purines (adenine, guanine) and pyrimidines (cytosine, thymine, uracil, as shown in Figure 7.1). Surprisingly, it was not until the 1940s that the crucial biological role of the nucleic acids began to be firmly suspected and not until the 1950s that the structure of these substances became known in some detail. The work that led up to this remarkable achievement came from a pooling of information in three formerly separate areas of study: from the structurists who had spent much time prior to the late 1940s in determining the structure of proteins; from the biochemists who had

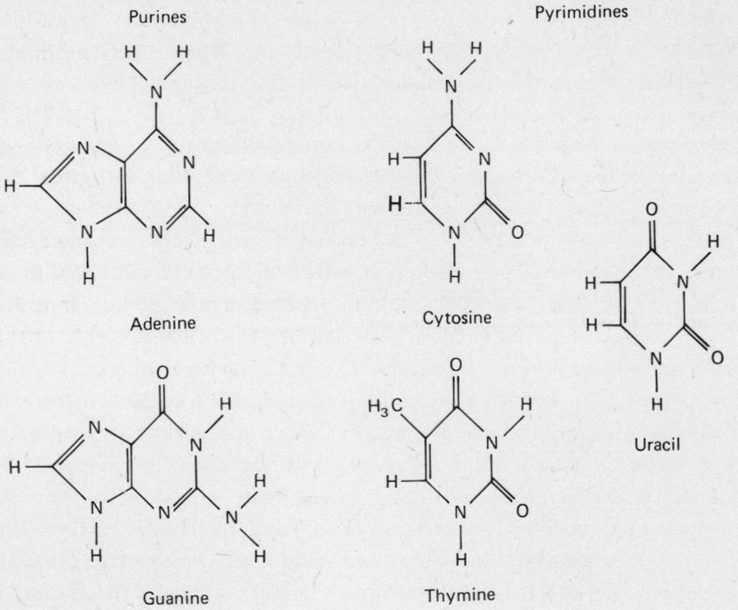

Figure 7-1

shown the connection between genes and the enzymes related to metabolic pathways; and from the informationists, who had implicated the nucleic acids as the carriers of hereditary information in at least a few groups of bacteria and viruses. While the chemical study of nucleic acids arrived late on the scene, many of the methods of determining molecular structure developed first in relation to protein became extremely important in deriving a model for nucleic acids that would account for their unique biological properties.

The Structurists's Approach

One of the most important of those techniques applied first to the study of protein structure, but later to that of nucleic acids was x-ray crystallography. Largely British in origin and development, the technique of x-ray crystallography became part and parcel of the structurists' school of thought. Proposed and developed around 1912 by H.W. Bragg (1862-1942) and his son W.L. Bragg (1890-1971), the technique of x-ray crystallography is relatively simple in conception.

Most crystals of any particular substance are relatively pure samples in which all the molecules (or atoms or ions) are arranged at definite and regular spatial intervals, forming a latticework of constant dimensions. In crystals, the molecules are all oriented the same direction in space, with their axes in parallel. If an x-ray beam is aimed at the crystal, the beam is deflected by the atoms or molecules in the crystalline latticework. Because of the regularity of spacing and of spatial orientation of the individual parts, the crystal acts like one giant molecule. The angle and nature of the x-ray diffraction is recorded on a photographic plate placed beyond the crystal from the x-ray source (see Figure 7.2). The diffraction pattern is different for every specific kind of protein. The pattern can be used to deduce the shapes of the molecules much as one might deduce the three-dimensional shape of a tinker toy structure by examining its shadow on a screen from several different angles. How the diffraction pattern, which is nothing more than a group of dots on a screen, can be translated into the tree-dimensional shape of molecules in the crystal proved to be the Gordian knot of x-ray crystallography.

Using crystals of small molecules such as rock salt and diamond, the Braggs had successfully deduced molecular structure, at the same time developing the techniques necessary for analysis of x-ray diffraction patterns. The methods were then extended to larger molecules such as the protein of hair. Even as the principles of x-ray diffraction began to be worked out,

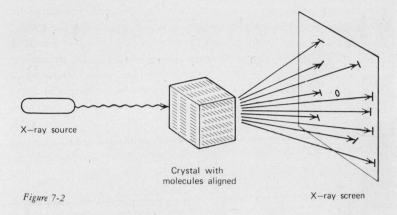

X—ray source

Crystal with
molecules aligned

Figure 7-2

X—ray screen

however, one of the major drawbacks was the tedium involved in analyzing all of the permutations of spacing patterns on the diffraction plate. Despite this, John Kendrew and Max Perutz, in Bragg's laboratory (the Cavendish Lab at Cambridge), succeeded in deriving most of the detailed structure for two related proteins, hemoglobin and myoglobin,[1] by the early 1960s. The molecular structures that Perutz and Kendrew elucidated showed not only the immense complexity of protein molecules but, more important, showed how a knowledge of three-dimensional structure could illuminate function. For example, Perutz and Kendrew could show that the "active site" of the hemoglobin molecule, a flat ring of atoms known as the "heme group," where the oxygen attaches, was tucked away in a fold on the surface of the molecule (Figure 7.3) surrounded by water-repelling side groups of atoms. This physical structure greatly facilitated the binding of oxygen with the active heme group.

Perutz and Kendrew's work on proteins was highly significant for subsequent developments in the study of the nucleic acids. Their conclusions emphasized the importance of studying three-dimensional structure as a key to understanding molecular function. Such findings stimulated those who were trying to investigate analogous problems in nucleic acids. Probably more important, the work on proteins elucidated many new techniques of analysis which simplified and refined the process of x-ray diffraction. While Perutz and Kendrew were working on proteins at Cambridge, another

---

[1] Hemoglobin is the main oxygen-carrying molecule in red blood cells; myoglobin is a very similarly shaped molecule that serves as the oxygen-carrying molecule in muscle cells.

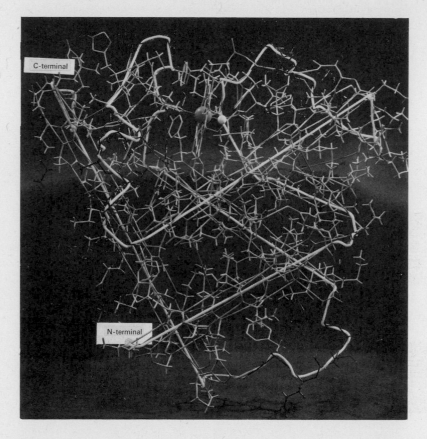

Figure 7-3   A photography of the Three-dimensional model of myoglobin (from a print by J. C. Kendrew. By permission of J. C. Kendrew.)

group, consisting chiefly of Maurice Wilkins (1916-   ) and Rosalind Franklin (1920-1958), was also using x-ray crystallography to study the nucleic acids at Kings College, London. The two groups had much contact and shared many of their problems and findings, especially with regard to technical developments.

By virtue of their backgrounds and inclinations, most x-ray crystallographers are highly empirical in approaching their work. Although not unconcerned about the biological implications of studying proteins or nucleic acids, their primary interest and motivation seems to have been molecular configuration. They were not prone to large-scale theorizing or to flights of imagination. In a very sober and rational way they stuck to the facts and developed general models of their molecules hesitatingly and with some caution. By contrast, the work of the informationists, beginning in the middle and the late 1930s, must have appeared rash and highly speculative. The informationists' goals were modestly summed up by one of their number as "a quest for the secret of life." Their approach was to unravel the method of information transfer in the hereditary process—an aim that was ambitious and romantic to some, utopian and even bordering on vitalism to others.

The Origin of The Informationist School

The informationist approach to heredity grew out of a climate of thought that was greatly conditioned by the quantum theory. In fact, the influence was direct and explicit, coming initially through the physicist Niels Bohr (1885-1962), who was a poineer in the quantum field and a curious thinker on matters biological. In a lecture of 1932 titled "Life and Light," Bohr tried to point out that the attempt to answer the question "what is life," by reducing the organism simply to chemical interactions faced the same difficulties as the attempt to describe the atom by picturing the position of every electron. There was a kind of biological indeterminacy, he argued, since the conditions necessary to keep an organism alive (in order to study life) precluded breaking it down into its component chemical systems. To do so rendered it no longer alive and thus a different system.

Bohr went on to argue that his views did not throw rational biology out of the window and open the door for neo-vitalism. If the supposed irrationality of quantum physics were examined, he said, it became apparent that it was not irrational at all but had reached a new order of rationality, where statistical probability supplanted one-to-one cause and effect relationships. The same was also true, he thought, of biology. To accept the existence of life as an elementary fact, a given, for which no satisfactory explanation in the older sense of the word (i.e., a single mechanism) was ever possible, was not to embrace mysticism and vitalism. Biology, like physics, he felt, could move to higher levels of understanding as it employed new concepts and approaches that did not require that the only satisfactory explanation of living phenomena was a mechanistic and reductivist one.

Bohr's ideas found a fertile growing place in the mind of his student, Max Delbrück (1906-   ). Born in Berlin, Delbrück studied atomic physics in Gottingen in the late 1920s and for two years was a Rockefeller Foundation Fellow in physics in Bohr's lab in Copenhagen. In the mid-1930s he went to work with the biology group at the California Institute of Technology. At Cal Tech Delbrück came into close contact with Morgan and the *Drosophila* geneticists who had moved their base of operation from New York to Pasadena in 1928. Morgan and his students were convinced that the methods of classical physics and chemistry would ultimately account for the processes by which genes carried out their metabolic roles. But Delbrück saw from the first that some other apporach than old-style mechanism might well be necessary to move classical genetics beyond the gross, chromosomal stages. Classical physics and chemistry, the physical views adopted by Morgan and his followers, could not provide a complete and sufficient picture of how genes function.

To Delbrück genes could not be thought of as molecules in the same way that traditional physicists and chemists think of molecules. The latter group, he pointed out, sees chemical reactions as uniform processes, where components collide with each other randomly in accordance with the principles of chemical kinetics. On the other hand, chemical reactions in cells are highly specific and are often kept separate from each other although taking place in close spatial proximity, a situation totally unlike that which prevails in the reaction systems studied by most chemists. Genes exist in only one or two copies within a cell, yet each carries out a multitude of reactions over the course of time. Furthermore, genes as molecules are highly stable in a thermodynamic sense, resisting degradation (breakdown to simpler forms) in a manner highly unusual for their size and compexity. All of these factors seemed to Delbruck to be highly anomalous if one looked at genes in the simple manner of the physicist or chemist. Thus, while genes were undoubtedly molecules that could be described perhaps as a chemist describes any large polymer, a molecular description would not give any hint about how the gene achieved its special properties.

The ideas of Bohr and Delbrück led to the highly unusual notion (especially to physicists) that the study of biological phenomena might lead to the elucidation of news law in physics and chemistry. This idea was embodied in an influential book published in 1945, immediately after the conclusion of World War II, by the physicist Erwin Schrödinger (1887-1961), titled, *What is Life?* Schrödinger argued that the inability of classical physics and chemistry to account for phenomena of life did not mean that these sciences would never be able to offer any help in solving biological problems. In fact, it meant just the opposite: the study of life could possibly open wholly new vistas of the natural world that have been hidden to physicists by the study of strictly inorganic phenomena. As classical physics had to revise its explanatory criteria to account for quantum phenomena, so it might have to revise further its criteria to account for biological. Schrödinger was aware that his ideas possibly implied that life followed an irrational pattern akin to the old vitalism; he thus made a special point of showing that the major problem of biology or physics was not the question of whether organisms defied the laws of thermodynamics by creating, for example, more energy than they took in; it was obvious that they did not. The real problem was that of information transfer: how information was coded, how it remained stable during numerous transfers from one cell generation to another, and how occasional variation was introduced. There were no counterparts to these phenomena anywhere in the inorganic world of the physicist or chemist.

Schrödinger went on to modify an older idea of Delbrück's that genes preserve their structure from generation to generation because the chromosome, of which genes are parts, was structured like an aperiodic crystal. It was composed of numerous smaller, isomeric (similar in form) units, the exact nature and succession of which determined the hereditary code. Crystals, by virtue of their lattice structure, have considerable stability (all atoms or molecules are bound to all others arranged around them) yet, by the specific nature of the arrangement of similarly structured by not identical parts could contain genetic information. Schrödinger concluded that "living matter, while not eluding the laws of physics as established up to date, is likely to involve hitherto unknown 'other laws of physics' which, however, once they have been revealed, will form just as integral a part of this science [physics] as the former." A new physics was to emerge from the study of biology, just as it had emerged from the study of quantum phenomena.

Schrödinger's biological views had a considerable impact in the 1940s and 1950s, especially on a younger generation of physicists. Physics was suffering from a general professional *malaise* at the end of World War II. The horrendous spectre of atomic war and the destructive aspects to which physics could lead caused many physicists to reexamine the relations between their work and the benefit it could (or could not) bring to human welfare. In addition, a number of centers of physics research in central Europe (especially Germany) had been disbanded as a result of the Nazi movement and the war impact itself. Numerous German physicists were Jewish and thus had fled the Nazi regime, some emigrating to Britain, others to the United States. Germany, once the center of quantum physics, became a scientific wasteland under Hitler and in the immediate postwar years. The disaffection of many physicists from their traditional subject grew out of the feeling (even before the war began) that the major developments in quantum theory had already taken place. The remaining work to be done was the necessary but often less exciting tasks of working out implications of the theory or cleaning up its details. And finally, of course, there was the influence of Schrödinger himself. As one of the founders of the quantum theory, his shift of interest to biology was significant and caused a number of physicists to take notice. Attracted by the possiblilty of finding "other laws of physics," Schrödinger's approach led some physicists into what appeared to be the new frontier of scientific investigation. It does not take a psychological analysis to see how such an approach could attract a number of inventive, even romantically inclined physicists who had come to feel that the only work remaining in their field was that of clarifying details.

The Development of the Biochemical School of Genetics

Soon after the rediscovery of Mendel's laws one of the most frequently asked questions was how do genes function to control specific characters? Between 1905 and 1925 a number of workers, including H.J. Muller, Sewell Wright, J.B.S. Haldane, and Richard Goldschmidt, suggested that genes acted in some way by controlling cellular metabolism. Yet no one had any concrete means of investigating this problem, since neither biochemical tools nor means of genetic analysis were sufficiently advanced to make experimentation practical. However, as early as 1908, Archibald Garrod delivered the Croonian Lectures (sponsored by the Royal Society of London) on the topic, "Inborn Errors of Metabolism" (published as a series under this title in 1909). Garrod's studies indirectly demonstrated the strong possibility that Mendelian genes functioned by affecting specific steps in metabolic pathways.

In his clinical work, Garrod encountered four distinct disorders that he termed "metabolic diseases." These conditions were characterized by the inability of affected persons to completely metabolize (break down or build up) some substance; the incompletely metabolized intermediate products were excreted in the urine, a means by which the disorder could be detected. For example, the normal metabolism of the amino acid phenylalanine involves a series of steps, ending up with fumaric and acetoacetic acids, as shown below (in abbreviated form):

One disorder that Garrod encountered was know as alkaptonuria, or phenyl-ketonuria, in which this series of reactions is blocked at the site marked "X" above. The block is a result, Garrod found, of the absence or presence in an ineffective form, of the enzyme, homogentistic acid oxidase. Because of this block, homogentistic acid cannot be converted into other substances along the pathway and accumulates in the body, eventually being excreted in the urine. Now Garrod asked a fundamental question: "What causes this block to occur?" With the help of William Bateson, he studied the family histories of those individuals afflicted with alkaptonuria. The relationship seemed clear: the disease was not caused by germs or acquired at random by some general malfunction. The disorder was inherited and appeared to follow the pattern characteristic of a Mendelian recessive. This striking finding suggested immediately to Garrod that Mendel's genes might in some way affect specific chemical products in the body's biochemical pathways. In 1914 Garrod thought he had found a proof of his conception; measurements made by a colleague showed that in normal blood an enzyme could be isolated that was capable of oxidizing homogentistic acid, while in afflicted persons no such enzyme was found. However, these findings were never confirmed; only as recently as 1958 has it been clearly determined that in an affected individual the enzyme is completely absent in at least one kind of tissue: liver.

Garrod's work suggested the explicit relationship that might exist between a specifically altered gene (mutation) and a known block in a metabolic pathway. Garrod's work, like Mendel's, was largely ignored for 30 years. Why no one picked up Garrod's lead is one of those difficult historical questions that deserve more study. The idea was not so farfetched that it could easily be dismissed, since many other workers had developed similar hypotheses. Garrod had evidence for each of the four metabolic disorders he studied, suggesting that each was genetic in origin and related to a specific metabolic block. Perhaps the fact that he was studying a disorder in human subjects, where experimental proof was hard to obtain, may have detracted from Garrod's impact. And no other favorable organisms seemed available on which studies of genetic blocks of metabolism could be pursued more extensively and rigorously.

The question of a biochemical basis for genetics did not remain an obscure issue, despite the fact that Garrod's far-sighted suggestions did not seem important at the time. Almost from the outset of the classical genetics work, the implication was clear that genes in some way must bring about their effects by influencing or regulating the production of chemical substances. Among Morgan's group, for example, the question of how genes function was never absent from evening discussions in the laboratory. H.J. Muller, one of the youngest and most imaginative of the group, wrote in the draft of a textbook on heredity that was never completed:

". . . there is strong reason for believing (1) that different parts of 'loci' of the chromatin differ from one another and from corresponding parts in other species or varieties in regard to the composition or structure of the substance they contain, each minute portion or 'locus' thus having its own individuality; (2) that they are able to grow, by their activity in some way transforming other substances, while they add to themselves, into material specifically like themselves, while, conversely, in their absence substances like themselves cannot be formed. The retention by the loci of their own particular character or individuality, throughout the growth and the various activities of the cell, means (3) that the substances in them are influenced little, if at all, by the processes and the special peculiarities of the cell around them, but (4) that these various substances [of the chromatin], each in its own way, influence profoundly the structure and activities of the cell."

Muller's summary of the chemical characteristics of "loci" was written in 1912. It reflects the clear recognition shared by many of the early workers in genetics that genes in some way affect cellular processes at the chemical level. In ensuing years there was an increasing tendency among some geneticists to think of genes in terms of the new concepts of proteins and enzymes. The close connection between genes and the control of metabolism was suggested by a number of workers about the same time: for instance, by Oscar Riddle in 1909 and Jacques Loeb in 1915; and, in 1917, Richard Goldschmidt and L.T. Troland proposed that genes were themselves enzymes. The idea that genes somehow affect the production or function of enzymes was also suggested by Garrod himself in 1909 and by Muller in 1922.

The problem was, of course, that no one could do anything more than hypothesize about these possibilities; there was no way to experimentally distinguish one scheme from another. Classical geneticists headed largely by T.H. Morgan and R.A. Emerson recognized this difficulty and consequently concentrated their energies on the transmission problem alone, where experimental techniques were available. But a younger group of workers, perhaps because transmission genetics was no longer the exciting field it had been between 1910 and 1920, felt more strongly about attacking the problem of gene action. One of the first young workers to become interested in these problems and to seek an experimental approach was a Nebraska farm boy, George W. Beadle (1903-   ).

While still an undergraduate at the University of Nebraska, Beadle was employed as a research assistant to Professor F.D. Keim, who had just returned from Cornell (in 1922) where he had completed a Ph.D. thesis on the genetics of wheat hybrids. In order to prepare himself better for his job, Beadle read one of the current genetics textbooks, which fascinated him. At Keim's insistance, Beadle eventually went to Cornell to pursue graduate study instead of returning to the Nebraska farm as he had originally intended. At Cornell, Beadle was tremendously excited by the research atmosphere generated by Emerson and by the vibrant atmosphere created by the group of graduate students assembeld there. When he eventually undertook research for his own thesis, Beadle chose a topic in classical plant genetics: the genetic mechanism determining pollen sterility in corn.

In 1928, while engrossed in his research at Cornell, Beadle attended a seminar given by Bernard O. Dodge of the New York Botanical Garden. Dodge was working with the mould *Neurospora* and had observed some strange segregation patterns among offspring from a cross between two strains. Beadle and several of his friends had just been reading Bridges' account of certain unusual types of chromosome crossing-over in *Drosophila* and suggested the same mechanism might operate in *Neurospora*. Nothing came directly from this suggestion except that it introduced Beadle to a new organism, one that could be grown easily in the laboratory and that was highly favorable for studies in heredity.

Dodge had already tried to persuade T.H. Morgan that *Neurospora* was a favorable organism for genetic work, but Morgan had not been readily convinced. Eventually, when Morgan went to Cal Tech in 1928, he took a culture of *Neurospora* with him and continued culturing it in California. A graduate student of Morgan's at Cal Tech had been assigned to work on the mold and had experiments in full swing by 1931 when Beadle arrived in Pasadena as a postdoctoral fellow. Beadle was not at first particularly attracted to the *Neurospora* work, being more interested in pursing the cytogenetics he had learned at Cornell (with plants) but now applied to *Drosophila*. In 1933 Boris Ephrussi (1901-   ) came to Cal Tech as a Research Fellow of the Rockefeller Foundation to study the relationship between genetics and embryology. Although Morgan was extremely interested in this subject, he had made little headway in treating the problem experimentally. The reason was that those organisms most favorable to genetic research (*Drosophila*, or corn) were least favorable for embryological research; similarly, those most favorable for embryological work (sea urchins, amphibians) proved unfavorable for genetic work. Ephrussi and Beadle were not satisfied with the attempts that had been made to bridge the gap between genetics and embryology and they agreed to spend a year together trying to work on the embryology of *Drosophila*. This they did at Ephrussi's laboratory in Paris during 1935.

The essence of Beadle and Ephrussi's work was to try and find some way of studying the embryonic development of traits in *Drosophila* whose genetic patterns (and chromosomal linkages) were already known. They attempted, initially, to culture *Drosophila* tissues (adult, larval, etc.) *in vitro*, in order to study the effects of different substances on the development of specific traits. Beadle describes his initial attempts and subsequent lucky break.

"I arrived in Paris in May, 1935, and we immediately began to attempt to culture Drosophila tissues. This proved technically difficult; so, on Ephrussi's suggestion, we shift to transplantation of larval embryonic buds destined to become adult organs. We sought advice from Professor Ch. Perez of the Sorbonne, who was a widely recognized authority on metamorphosis in flies. He said we had selected one of the worst possible organisms and that his advice was to go back and forget it. But we were stubborn and before many weeks had devised a successful method of transplantation. The first transplant to develop fully was an eye. It was the occasion for much rejoicing and celebration at a nearby cafe."

Using this technique, Ephrussi and Beadle transplanted eye buds from *Drosophila* of one genetic constitution into the body of larvae of different genetic constitutions. The resulting color of the eye transplant varied according to the genetic make-up of both bud and host. The development of pigment molecules appeared to be dependent not only on the genes present in the transplant, but also on substances made available to the eye bud from its surroundings (i.e., the rest of the body in which it was residing). By making a number of transplants of many different kinds, Ephrussi and Beadle were able to devise a simple biochemical scheme in which individual steps in the synthesis of eye color pigment could be identified with specific intermediate substances, as shown below.

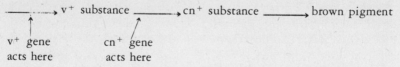

$$\longrightarrow v^+ \text{ substance} \longrightarrow cn^+ \text{ substance} \longrightarrow \text{brown pigment}$$

v^+ gene cn^+ gene
acts here   acts here

The v^+ gene controlled production of the v^+ substance, and the cn^+ gene controlled the production of cn   substance. Both genes were present in the wild-type *Drosophila*, yielding the brownish-red pigment, characteristic of wild-type eye color. If the cn^+ gene was mutated only v^+ substance was produced, the pathway being blocked beyond that point. If the v^+ gene, or both v^+ and cn^+ genes were mutated, the eye color was white. Now, if an eye bud from a *Drosophila* in which the v^+ gene was active (but the cn^+ gene was not) was transplanted into the body of a larva in which the cn^+ gene was active (and the v^+ gene inactive), the transplant developed wild-type eye color. These data suggested to Ephrussi and Beadle that genes controlled certain steps in the biochemical pathway for pigment production. Beadle and Ephrussi did not, however, hypothesize at the time how the genes effected this control.

There were problems with the Ephrussi-Beadle technique; it was cumbersome and, while it worked well for easily diagnosable traits such as eye color (where the pigment molecule was the end product), it did not prove useful for more complex traits where the end product was a structure instead of a chemical substance.

Beadle's work with Ephrussi had shown him that *Drosophila* was not the most favorable organism for studying how genes act. As with all higher organisms, *Drosophila* had two genes for every character, so that dominant genes often masked recessive ones. It was also a highly complex form, with myriads of metabolic pathways that could not be studied very directly. After returning to the United States, Beadle eventually took a job at Stanford University, where he met the microbiologist E.L. Tatum (1909-   ). Their association introduced a very fruitful era in the history of biochemical genetics.

Tatum's familiarity with microorganisms suggested the possibility of some simple form, such as a mold or bacterium, whose metabolic characteristics could be more easily studied. Together he and Beadle decided to work with *Neurospora*, largely as a result of Beadle's already established familiarity with its convenience for laboratory use. The advantages of Neurospora are many; chief among them are (1) relatively short generation time for producing sexual offspring; (2) ease of growth and maintenance in the lab (they grow in test tubes on a simple medium); (3) ease in identification of metabolic (i.e., biochemical) mutants; and (4) an adult stage that is haploid (only one set of chromosomes), thus allowing all mutant genes to show their phenotypic expression. In all these ways *Neurospora* was a far step ahead of *Drosophila* for the laboratory study of biochemical genetics.

Beadle and Tatum's basic experimental design was simple. They irradiated spores of normal *Neurospora* to increase the rate of mutation (although they could not induce any particular kinds of mutations). Since most mutations are deleterious, the experimenters expected that many irradiated spores would not be able to germinate on what was called "minimal medium." This medium contained the minimal raw materials and food that are necessary for normal (i.e., wild-type) *Neurospora* to live and grow. From the minimal raw materials provided, the organism can synthesize everything else it needs for normal metabolism. All irradiated spores were first grown on a "complete" medium, which contained all substances, both raw materials as well as those normally manufactured by the organism. After a good crop of each type of spore was harvested, they were then grown on minimal media to determine which stocks had mutations. Those that could not grow were hypothesized to have at least one metabolic block caused by mutation. Then spores from the stocks known to contain mutations were grown on a number of "modified media," each lacking a specific material (but containing all the others). For example, if the spores from stock A could grow on all the modified complete media except the one lacking Vitamin B, then it could be reasoned that stock A had a metabolic block in the pathway for the synthesis of this vitamin.

After identifying a number of such mutants, Beadle and Tatum did genetic analyses of spore formation in each strain. The results showed that the metabolic blocks were directly correlated with gene segregation. Thus it appeared that observable metabolic blocks are associated with genetic mutations. But more specifically, since each step in a metabolic pathway was known to be catalyzed by a specific enzyme, Beadle and Tatum concluded that gene mutations caused enzyme changes and that each gene must control the synthesis of one particular enzyme. From this work they developed what has been called the "one gene-one enzyme" hypothesis: the idea that each gene produced one specific enzyme. Beadle and Tatum pursued their work in complete ignorance of Garrod's similar findings 30 years previously. But, in reality, as they acknowledged, they only confirmed what their predecessor had stated so eloquently.

Today biologists tend to regard the one gene-one enzyme hypothesis as only part of the story, since it appears that a gene codes for a single poly-peptide chain, not for a whole enzyme or protein molecule. What is perhaps most important about Beadle and Tatum's work is that they found an organism in which gene function could be studied experimentally. *Neurospora* was not only simple to culture in the laboratory, but its spores were easy to analyze genetically and, through irradiation techniques, mutations could be produced quite easily. Most important, the organism normally has only one instead of two genes determining each character, thus there is no problem of recessive genes being masked by dominants; every gene shows its effects. The genotype of *Neurospora* can be tested for its metabolic characteristics directly, by whether it grows or does not grow on a specific medium. The introduction of *Neurospora* into biology did for biochemical genetics what *Drosophila* did for chromosomal genetics.

The function of genes as controllers of the structure of specific polypeptide chains received important confirmation in the decade following Beadle and

Tatum's work. One example will illustrate how ideas developed on this subject. While the purpose here is not to chronicle a list of discoveries, it is important for understanding the development of molecular genetics to see what kinds of ideas and evidence were available during its period of growth.

In 1949, J.V. Neel (1915-   ) showed that the human genetic disorder, sicklecell anemia, is inherited in a simple Mendelian fashion. In the same year, Linus Pauling (1901-   ) with several co-workers, showed that the hemoglobin protein of sickle-cell patients is somehow structurally different from normal hemoglobin (they distinguished the two by electrophoresis). From his knowledge of protein structure, Pauling reasoned that the difference must be due to some difference in amino acid composition. Some years later, in 1957, V.M. Ingram (1924-   ) did a detailed amino acid analysis of sickle-cell and normal hemoglobin and showed that the two differed by only a single amino acid. Yet this difference was enough to seriously affect the health of afflicted individuals (persons with sickle-cell anemia have less efficient means of carrying oxygen to their tissues under conditions of low oxygen pressure in the blood). That sickle-cell anemia was the result of a single Mendelian mutation suggested that genes somehow determined the sequence of amino acids in a polypeptide chain. Thus, by the late 1940s and early 1950s, it was becoming 'apparent that genes controlled cellular metabolism by controlling the production of specific proteins. Proteins, it seemed, were the most direct phenotypes produced by an organism's genotype.

Development of the Unified Informationist Approach

The romantic phase of the informationist school can be thought of as beginning with the introduction of bacteriophage (viruses that attack bacterial cells) as a research "tool" by Max Delbrück in 1938. It ended roughly around 1952 with the definitive work of A.D. Hershey and Martha Chase showing the DNA was identifiable as the hereditary material. The romantic phase developed out of the search for the physical basis of heredity, a topic that Delbrück had come face to face with in Morgan's laboratory in the late 1930s. Morgan and his co-workers had tried to calculate the size of various gene-units and had often raised the question of what components of the chromosome carried the genetic message. But *Drosophila* was not a favorable organism for answering such questions—nor did it even suggest a concrete question that could be answered experimentally. It was for this reason that Delbrück decided to work not with such complex and multi-celled forms, but with bacteriophage, as "ideal objects for the study of self-replication." He soon met Salvador Luria (1912-   ), a refugee from fascist Italy, and A. D. Hersey (1908-   ), and together they formed the beginnings of the famed "phage group."

We know today that bacteriophage, like all viruses, consist of an outer coat of protein that surrounds nucleic acid (usually DNA, but in some forms ribonucleic acid, RNA). Each phage particle is shaped something like a syringe (see Figure 7.4) and attaches its smaller end to the outer coat of bacterial cells. It then "injects" its core of nucleic acid (in the case of phage it is DNA) into the bacterial cell, leaving the coat attached to the outer cell wall. Once inside the bacterium, the phage nucleic acid commandeers the cellular machinery of the bacterium for producing not bacterial parts but viral parts. These parts, consisting of new DNA (which has been replicated) and new protein for the coat, are then assembled into 50 to 100 new phage, which erupt from the cell and can go on to infect other bacteria.

The phage group was prophetic in singling out viruses as favorable systems for answering questions such as "What molecules transmit genetic information?" Viruses have a number of special advantages that have allowed them to occupy vitually the same position in the history of molecular genetics that *Drosophila* occupied in the history of classical genetics. For one thing, bacteriophage are easy to grow (they are plated out on a smooth layer of bacterial cells in a petri dish), and many millions can be cultured in a small space. For another, their generation time, like that of bacteria, is quite rapid—occupying anywhere from 20 to 30 minutes. Thus numerous generations can be obtained in relatively short periods of time. And last, because they consist of only two types of molecules, protein and nucleic acid, they are highly favorable for rigorous experiments that could follow the fate of either the nucleic acid or protein during the reproductive process.

The early phage workers described carefully the life cycle of viruses and elucidated the course of infection of host cells. They also showed that the genetic material of viruses undergoes mutation, like all higher forms, and that genetic recombination between different strains of the same virus is possible. Despite these very important advances, however, the early phage workers did not come to the solution of the major problems of gene structure and function. One reason was their persistent interset in the problem of information, which led them at the time to focus chiefly on proteins instead of nucleic acids as the probable genetic material. It was because proteins were the only molecules at the time known to be a linear sequence of specific units that they seemed to many informationists the most likely vehicle for information storage and transfer. Another reason was that many of the workers in

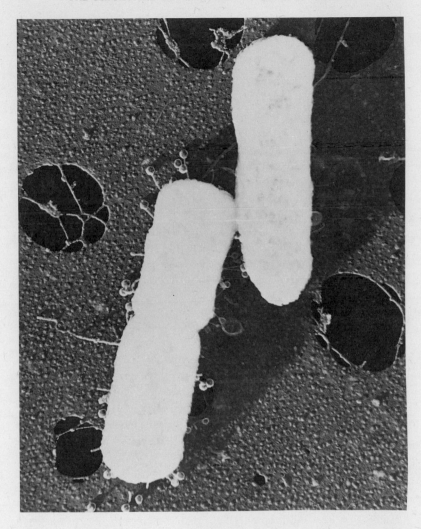

Figure 7-4   (From "Viruses and Genes," by Anderson, Jacob, Kellenberger, and Wollman, June 1961, Scientific American. By permission of the authors.)

the romantic phase were disdainful of contemporary biochemistry. They felt its methods were to imprecise to add anything significant to the study of how individual, single molecules (each gene was thought to exist in only one or two copies inside most cells) behave in biochemical reactions. Thus most of the phage workers ignored the work in biochemical genetics that was beginning to produce fruitful results in the early 1940s.

In 1944 O.T. Avery (1877-1955), Colin MacLeod (1909-   ), and Maclyn McCarty (1911-   ) published a highly significant, if cautious paper, regarding the so-called "transforming principle" discovered by the British microbiologist Fredrick Griffith (1877-1941) in 1928. Griffith had shown that if a living, benign strain of bacteria is injected into a host animal simultaneously with a nonliving (ususally heat-killed) virulent strain (the two must be of the same bacterial species and type), some of the benign forms become actively virulent. Avery, MacLeod, and McCarty set about to isolate the substance that they assumed must pass from the heat-killed into the living bacteria to transform the latter into virulent types. From the fact that the main bulk of chromosomal material was protein, they deduced that protein must the "transforming principle." Yet none of the protein fractions isolated from heat-killed cells and placed into cultures of benign bacteria produced any effect. However, a "biologically active fraction" was eventually isolated that turned out to be DNA. DNA seemed to be able to convert one genetic type of bacteria into another; it might well be, in the words of Avery, MacLeod, and McCarty, "the fundamental unit of the transforming principle." The authors were unwilling at the time to extend their findings to any general conclusions. They conceded that it was "of course" possible that "the biological activity of the substance described is not an inherent property of the nucleic acid but is due to minute amounts of some other substance absorbed to it or so intimately associated with it as to escape detection." Even if this were not the case, they noted, their results would only support the conclusion that DNA was the carrier of a particular hereditary trait in a single type of bacterium; there was no evidence that all traits in all organisms were transmitted by the same molecule. Avery and his co-workers were, as one historian has remarked, "almost neurotically reluctant to claim that DNA was genes and genes were simply DNA."

The reason for this reluctance was not hard to surmise. Between 1910 and 1940 the classical Mendelian theory developed completely apart from biochemisty. And, as we have seen, the later union developed in part by Beadle, Ephrussi, and Tatum dealt not with the biochemical nature of genes themselves, but with the metabolic sequences controlled by genes. The bridge between the two seemed impossibly remote to many in the early 1940s.

The romantic phase of the informationist school came to a climax in the early 1950s with the definitive studies of A.D. Hershey (1908-   ) and Martha Chase (1927-   ) using radioactive tracers to follow the molecular events in phage infection. Motivated partly by a desire to eliminate the circumspection surrounding the conclusions of Avery, MacLeod, and

McCarty, they labeled the DNA of phage with radioactive phosphorus and the protein coat with radioactive sulfur. Their results, published in 1952, showed that when a phage infects a bacterial cell, it injects its DNA into the host, leaving the protein coat on the outside. Subsequent studies showed that only the phage DNA, not the protein, was involved in the biochemical events involved with the replication of new phage particles. This demonstration presaged the end of the romantic era among informationists because it showed that in the study of phage, no new paradoxes might crop up and come into sharper focus. No new laws of physics or biological organization seemed to be forthcoming. The basic problem of reproduction could now be restated in terms of two functions of phage DNA: autocatalysis and heterocatalysis.

Autocatalysis was a process, long recognized in biochemistry, whereby a molecule can catalyze some of the early steps in its own formation. Heterocatalysis was the process whereby a molecule catalyzed the formation (or breakdown) of other different molecules. By means of the former, phage DNA was seen to replicate itself many times over to make many copies of viral genetic instructions. By the latter, phage DNA induces in some way the synthesis of virus-specific proteins that govern the reactions involved in phage growth and assembly. In 1952 no one was certain how either of these functions were carried out, or exactly what the properties of such a molecule as DNA must be in order to engage in two such different sorts of events. Indeed this question became the subject of the second, or what has been called by one phage worker the dogmatic, phase of the informationist school, lasting from approximately 1950 until the middle 1960s. It was during this period that workers motivated by the earlier informationist approach brought together the results from biochemical genetics and from the structurists' studies of macromolecules to deduce the actual structure of DNA and from that infer a number of its functional properties. The dogmatic phase was dominated by the work of James D. Watson (1928-   ) and Francis Crick (1916-   ).

At the University of Chicago, where he was an undergraduate, J.D. Watson had been hypnotized by Erwin Schrödinger's book, *What is Life?* He later reported that this book "polarized" him toward "finding out the secret of the gene." Having been turned down by two graduate schools, Harvard and Cal Tech, he went to Indiana University, whose biology department was graced in genetics by the presence of H.J. Muller, recent winner of the Nobel Prize (1947) for his work on x-ray induction of mutation. Having decided quickly that Muller and the *Drosophila* school offered little hope for answering Schrödinger's challenge, Watson took a doctorate in phage genetics under Salvador Luria, who had come to Indiana in 1943. Watson was also stimulated by Luria's friend and collaborator Max Delbrück, whose role in catalyzing Schrödinger's thinking on biological matters had already been acknowledged. Watson later stated that he chose to work with Luria partly because the latter was known to be Delbrück's associate in the path-breaking phage work. According to reports, from the time Watson met Delbrück in Luria's apartment in Bloomington in the spring of 1948 onward, he was "hooked" on the aims and methods of the phage school.

From Delbrück Watson imbibed some of the early informationist's open disdain for classical geneticists, who were seen as "nibbling around the edges of heredity rather than trying for a bull's eye [i.e., to discover the nature of the hereditary molecule and its means of auto and heterocatalysis]." But he was less swayed by Delbrück's disregard for biochemistry. Like his teacher, Luria, Watson felt that if one wanted to know how genes behaved, it would be useful to know what they were made of. It was with these ideas in mind that Watson went off to Europe on a series of postdoctoral fellowships that eventually landed him at the Cavendish Laboratory in Cambridge in the spring of 1951. It was here that by chance he met Francis Crick and began one of the most productive and exciting, if enigmatic, collaborations in the history of modern biology.

Francis Crick attended University College, London, taking a degree in physics in the late 1930s. He continued there as a graduate under the guidance of Professor E.N. da C. Andrade (1887-   ), an accomplished expositor of the Bohr theory of the atom. Crick's doctoral work at first consisted of studies on the viscosity of water at high temperatures (above 100°C); but when World War II came along, the laboratory was closed and Crick, along with many scientists and students, joined the Admiralty Research Laboratory at Teddington. Throughout the war he worked closely with various teams on circuit-operating systems that would detect and detonate enemy mines. Even after peace returned he remained at the Admiralty with the ultimate intent of going into fundamental research either in particle physics or physics applied to biology. For the latter inspiration he was indebted to Schrödinger's *What is Life?*, which he, too, had read sometime near the close of the war.

What had most impressed Crick about Schrödinger's book was not its romantic call to seek new laws of physics but the fact that "fundamental biological problems could be thought about in precise terms, using the concepts of physics and chemistry." After reading Schrödinger's book, he wrote that one is left with the impression that "great things were just around the corner." One of those "great things" to Crick was J. D. Bernal's (1901-

1971) recent x-ray crystallography studies on proteins and nucleic acids, and he applied in 1946 to work with Bernal's group at Birbeck College, London. Rejected by Bernal, however, he eventually obtained an appointment at Cambridge, ultimately with a small research unit at Gonville and Caius College. Here he came to work with Max Perutz on the x-ray crystallography of hemoglobin. Although Crick knew nothing of x-ray crystallography prior to his work with Perutz, he taught himself much of the subject and read through Perutz' and Kendrew's early studies on the structure of hemoglobin and myoglobin.

After a year at Cambridge, Crick presented a seminar titled "What Mad Pursuits" in which he challenged some of the basic techniques and modes of interpretation that Perutz and Kendrew had been following in their x-ray crystallography work. He felt that the hemoglobin molecule was much more complex than the hatbox model that Perutz and Kendrew were accustomed to seeing from their data. Since the molecule was so complicated, and the methods not yet refined enough to show all the little kinks, short, straight chains, and other meanderings of the polypetide chain, oversimplified pictures were not unexpected. But, as Crick pointed out,

"It is one of the occupational hazards of the sort of crystallography in which you do not get results within a reasonable time, that those who work in it tend to deceive themselves after a bit; they get hold of an idea or an interpretation and unless there is someone there to knock it out of them, they go on along those lines, and I think that was the state of the subject when I went to the Cavendish."

Unfettered by a traditional education in crystallography and, in many ways naive of its great complexity, Crick set about to knock out of the Cavendish group some old habits of thinking about molecular structure.

It was perhaps as much Crick's habit of talking and thinking almost continually, as his bold and iconoclastic challenges to established ideas, that attracted young Watson to him shortly after arriving in Cambridge in 1951. After two years wandering through several uninspiring biochemical laboratories on his postdoctoral fellowship, Watson finally found in Crick someone who shared his same interests—the molecular nature of the gene. The intellectual stimulation was immediate and catalytic. As Watson wrote in a letter to Delbrück in December 1951:

"[Crick] is no doubt the brightest person I have ever worked with and the nearest approach to Pauling I've ever seen—in fact he looks considerably like Pauling. He never stops talking or thinking, and since I spend much of my time in his house (he has a very charming French wife who is an excellent cook) I find myself in a state of suspended animation. Francis has attracted around him most of the interesting young scientists and so at tea at his house I'm liable to meet many of the Cambridge characters."

With their common intellectual ancestry in Schrödinger, Watson and Crick approached biology from similar points of view, if not with similar backgrounds. Crick was essentially a physicist with little or no interest in biochemistry. Watson was a phage geneticist who had come to recognize that any answer to the question of how genes worked would require some knowledge about cell chemistry. Crick's interest in genetics had brought him to work on biologically important macromolecules, which at that time meant to him proteins. Watson's interest in genetics led him to the structure of the molecules of heredity, which at that time meant to him primarily the nucleic acids. The origin of this difference in viewpoint and the means of its resolution, provides the key for understanding how Watson and Crick decided to focus their efforts on DNA as the hereditary material. They began their work a year or more before the publication of Hershey and Chase's definitive proof that DNA was indeed the hereditary material. Hence, their reasons for choosing to study DNA rather than protein depended upon other influences.

From the late 1940s on, Crick had been searching for some way to link gene structure to protein structure. He was convinced that the specificity of proteins lay in their amino acid sequences and that this sequence must somehow be related to the arrangement of the hereditary material. But he was not certain what the hereditary material was, and he somehow did not feel compelled to grant the nucleic acids much of a role in carrying hereditary information. Crick was, in fact, a close friend of Maurice Wilkins, who had from the mid-1940s on, been studying the x-ray diffraction patterns of DNA at Kings College, London. Although Crick had heard Wilkins lecture on DNA on more than one occasion, he seems not to have been impressed with the idea either that DNA was an important molecule to study or that Wilkins' preliminary data suggesting that the molecule was helical in shape was very interesting. On one occasion at tea Crick is reported to have quipped to Wilkins, "What you ought to do is to get yourself a good protein!" DNA was, in fact, very difficult ot study by x-ray diffraction, since preparation of single crystals is extemely difficult, if not impossible. Diffraction patterns are thus considerably more difficult to interpret. Crick was right in one way—no one in their right mind would choose on *a priori* grounds to study DNA by x-ray diffraction analysis, but Wilkins and others had some sense of the molecule's importance that Crick could not grasp. This is where Watson entered the picture.

Watson was, of course, familiar with the work of Avery and his co-workers from 1944—by that work was inconclusive and did not in itself lead him directly toward DNA as the hereditary material. Probably more important was Luria's general hunch that DNA was "beginning to smell like the essential genetic material." In fact, primarily on this basis, Luria had encouraged Watson to go for his postdoctoral work to the biochemical lab of Hermann Kalckar (1908-   ) in Copenhagen to study the chemistry of DNA. Although Watson was visibly bored by these chemical studies, in the spring of 1951 he did become acquainted with Wilkins' work at Kings College. Wilkins' major focus seemed to Watson to have been the structure of the molecule itself; he was not disposed in his structural theories to consider biological (genetic) function in any very real ways. Wilkins was primarily a structurist at heart. Yet Watson's move to England to learn more about the x-ray studies had the unexpected bonus of a meeting with Crick— and that turned the tide.

In Watson, Crick found a biologist steeped in genetics (which Crick was not), but who was anxious to find out how the gene worked at the molecular level. In Crick, Watson found a physicist who not only knew something about x-ray crystallography, but who was primarily interested in relating gene structure to biological function. It was something new and invigorating to Watson, who felt that the other individual efforts he had seen— Kalckar's group or even Wilkins'—were by themselves leading in uncertain and possibly unfruitful directions.

The details of Watson's and Crick's collaboration and the technical problems (or concepts) that they had to overcome to piece together the structure of the DNA molecule has been recounted in several recent publications, the most exciting and titillating of which is Watson's controversial book, *The Double Helix.* It would be beyond the scope of our present study to recount the chronicle of this discovery, yet a brief summary of Watson and Crick's work between the fall of 1951, and the publication of the historic paper in *Nature* in April 1953 is necessary to understand how the framework of molecular biology was built from the union of the structurist, informationist, and biochemical schools of genetics.

Once they had decided to work on the structure of DNA, Watson and Crick attempted to devise a molecular model that would not only be in agreement with x-ray diffraction data, but that would also account for the dual functions of auto-catalysis and heterocatalysis. From the x-ray diffraction studies of Wilkins and his associate Rosalind Franklin, it was clear that the DNA molecule consisted of stacked layers of subunits whose regularly spiralled geometric shape was repeated every 3.7 Å units. (See Figure 7.6).

It was also clear that the molecule was a long-chain polymer of some sort whose diameter was constant all along its length. As we saw above, the chemical composition of DNA had been determined by the early 1930s; it was shown to consist of four building blocks, two purine and two pyrimidine bases, linked to sugar and phosphate groups to form what are called nucleotides (see Figure 7.5). The sugar groups of one nucleotide were known to be chemically linked to the phosphate of another, producing a sugar-phosphate "backbone" that ran the length of the molecule. This was the basic information Watson and Crick had at their disposal to start with. Their aim was to determine how these parts fitted together in a structurally and biologically meaningful way.

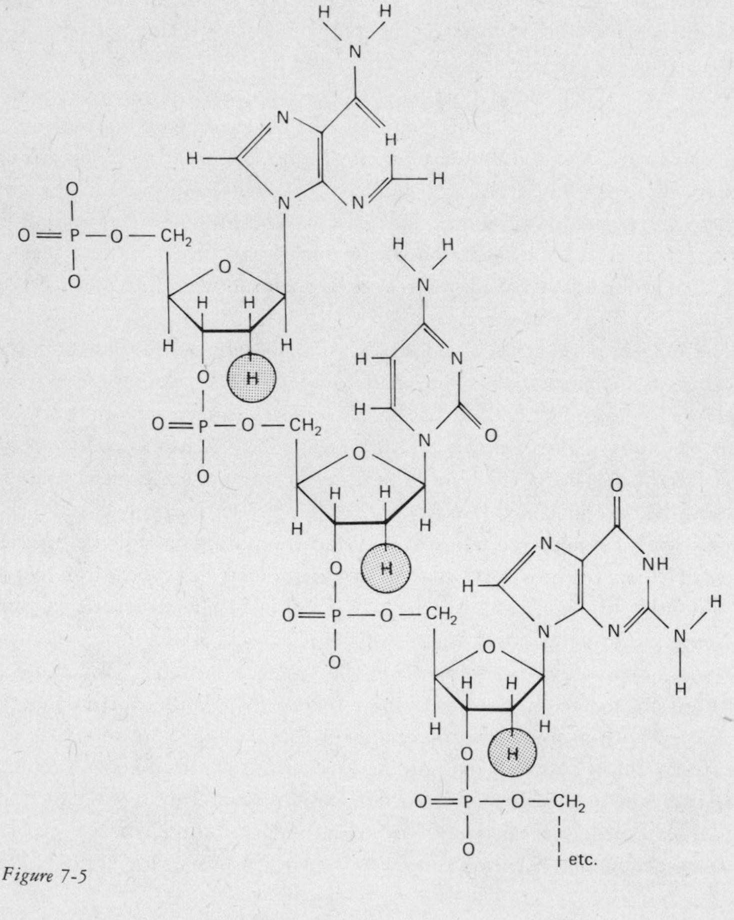

Figure 7-5

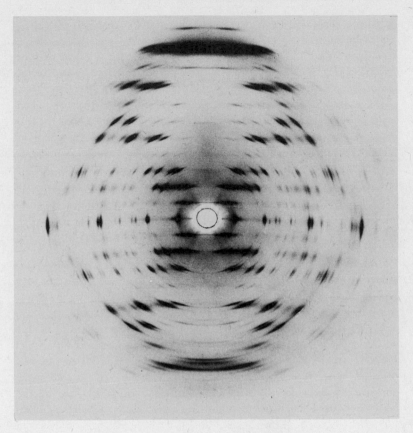

Figure 7-6   (By permission of M. H. F. Wilkins.)

The main problem can be stated as follows. Any arrangement of nucleotides in the DNA molecule had to account for the regularity of the molecule's structure, as well as its chemical stability. It also had to account for how the molecule could replicate itself faithfully. These considerations meant that one had to determine not only what physical arrangement of the parts was geometrically possible, but particularly what known interatomic and intramolecular forces could hold the parts together. The sugar-phosphate backbone was held together by strong forces known as covalent bonds, the more common types of chemical linkages between atoms in a molecule. But it seemed clear from x-ray diffraction data that each DNA molecule consisted of several (two, three, perhaps even more) sugar-phosphate backbones with their attached purine or pyrimidine bases. How

were these chains held together? One possibility was that the several nucleotide chains were bonded by weak attractive forces (such as hydrogen bonds
or van der Waals' forces) between like bases (i.e., adenine and adenine,
cytosine and cytosine), or between like base types (i.e., purine and purine,
pyrimidine and pyrimidine). Another was that unlike bases attracted: that a
purine attracted a pyrimidine. Still other possibilites were that parts of one
sugar-phosphate backbone attracted parts of others, or that bases on the same
molecule attracted each other, such that the single sugar-phosphate
backbone was bent and folded back on itself. Until well into 1951 this latter
was the type of structure that Crick thought most possible.

In 1951, however, three factors influenced the direction of Watson's and
Crick's thinking. One was a chance meeting with a young Cambridge
mathematician, John Griffith, who was also somewhat interested in certain
biochemical problems. Griffith agreed to calculate for Watson and Crick
what the attractive forces (weak interactions) would be between *like* bases in a
DNA molecule. Crick was still thinking in terms of bases on the same chain
attracting each other, when he met Griffith some time later at tea and asked
him if he had made the calculations. Griffith replied that he had, and that he
found from theoretical considerations that, instead of like bases attracting
unlike bases seemed to attract each other. Crick was not disturbed, because
he immediately saw another possiblity, complementary pairing, which
would explain replication very nicely: A goes to make B and B goes to make
A. Still, however, this finding did not illuminate the overall three-
dimensional nature of the molecule. More important was the meeting
between Watson and Crick and the Austrian refugee biochemist Erwin
Chargaff (1905-   ), then at Columbia University's College of Physicians
and Surgeons. Chargaff was visiting Cambridge in June of 1952 and was
introduced to Watson and Crick by Kendrew. According to Crick, the
following conversation took place at their meeting.

"We were saying to him as protein boys, 'Well, what has all this work on
nucleic acids led to? It has not told us anything we want to know.' Chargaff,
slightly on the defensive, replied: 'Well, of course, there are the 1:1 ratios.'
So I said: 'What is that?' And he said: 'Well, it is all published!' Of course, I
had never read the literature, so I would not know. Then he told me and the
effect was electric. That is why I remember it. I suddenly thought: 'Why,
my God, if you have complementary pairing, you are bound to get a 1:1
ratio. By that stage I had forgotten exactly what Griffith had told me. I did
not remember the names of the bases. Then I went to see Griffith and asked
him which his bases were and wrote them down. Then I had forgotten what

Chargaff had told me, so I had to go back and look at the literature. And to my astonishment the pairs that Griffith said were the pairs that Chargaff said."

Thus, in the spring of 1952, Crick realized that complementary pairing of unlike bases was likely to be the basic format on which DNA molecules were built. This realization was extremely important, because it disclosed how the molecule could be held together and how it could replicate itself. However, there was still no evidence as to how many strands the molecule consisted of or of their spatial relations to each other.

A third clue came not so much from the content but from the method introduced by Linus Pauling in his studies on the alpha-helix of protein. Pauling's theory, published in 1950, suggested that proteins of every polypeptide chain in most known proteins automatically fold into a coil, like a bedspring, held in place by hydrogen bonds between amino acids. While this actual idea of self-bonding between parts of the same molecule may have given rise to Crick's erroneous conception that the DNA molecule was perhaps a single strand folded back on itself, Pauling's method of approach was important. By working from theoretical considerations to model building and then testing the model against x-ray diffraction results, Pauling provided a unique physical way of viewing molecular dimensions. Adopting this approach, Watson and Crick used scale models of the purine and pyrimidine bases to determine what dimensions and arrangements would fit the theoretical requirements of hydrogen bonding and the empirical requirements of Chargaff's rules. This method proved invaluable, because it allowed Watson and Crick to propose various alternatives and test them out against x-ray diffraction patterns of DNA to see if any correlation could be found.

The work of Watson and Crick during the winter of 1953 became more and more frenzied. Their zeal to build an accurate model of DNA was whetted by the knowledge that Pauling also had promised to produce a model very soon. The alternatives had narrowed themselves down to a two- or three-chain molecule (not a singel chain) with the bases pointing inward and the sugar-phosphate backbone outward, or the bases pointing outward and the sugar-phosphate backbones inward. Wilkins and Rosaline Franklin had earlier claimed that the x-ray data tended to rule out a two-stranded structure, but they were not prepared at that time to publish their results, and so Watson and Crick did not have the data necessary to check the possibilities. Watson then visited the Kings College group in early February and discovered from Wilkins that the x-ray density patterns did not rule out

Figure 7-7   (By permission of J. D. Watson.)

the two-stranded model after all. Furthermore, he also learned that Rosalind Franklin's original suggestion of nearly a year earlier, that the sugar-phosphate backbone must be on the outside, was entirely verified by recent diffraction studies. The last two weeks in February turned out to be the critical period. While Crick was trying periodically to work on his thesis, Watson was building models of two- and three-stranded DNA molecules. He tried primarily to place the sugar-phosphate backbones on the inside of the molecule, because as long as the bases were on the outside, their exact chemical interactions could be conveniently ignored. But, if they were placed on the inside, according to Watson, "the frightful problem existed of how to pack together two or more chains with irregular sequences of bases." Neither Watson nor Crick in mid-February saw much hope of resolving this problem.

Organic molecules exist frequently in one of several structural forms, depending on various environmental conditions (such as pH). In working with their cardboard molecular models, Watson and Crick had used one of the several possible forms of the purine and pyrimidine bases. After trying in several ways to fit the bases together across the strands of the molecule, they sought the advice of Jerry Donohue (1920-   ), an American chemist who shared their office. Donohue pointed out that Watson and Crick had been using the wrong form of the molecule and, when he suggested the right form, the idea hit like a flash. Crick described the point that became suddenly very clear:

"Jerry and Jim were by the blackboard and I was by my desk, and we suddenly thought 'Well perhaps we could explain 1:1 ratios by pairing the bases.' It seemed too good to be true. So at this point [Friday, February 20] all three of us were in possession of the idea we should put the bases together and do the hydrogen bonding."

With the wrong form of the bases, hydrogen bonding had been difficult to conceive, since the spatial distances were such that these bonds could not form easily across the molecule. With the new form of the bases, however, hydrogen bonding appeared quite possible. Still, the question was, "Which bases bonded to which?" One day, as he played around with the cardboard models on his desk, Watson noted that an adenine-thymine pair held together by two hydrogen bonds had the same molecular shape and dimensions as a cytosine-guanine pair. This pairing scheme would account at one blow for both the constancy in diameter of the DNA molecule and for the 1:1 base ratios reported by Chargaff. For nearly a week Watson and Crick did nothing but build models of DNA using the A-T, C-G base-pairing scheme (see Figure 7.7), trying to show that every major angle of rotation shown by the x-ray diffraction data could be accounted for. Through various flukes and purposeful designs, including the use of Watson's excellent memory from his recent visit to Wilkins' laboratory, Watson and Crick were able to test their model against some empirical results. The two fit extraordinarily well.

The Watson-Crick model of DNA, announced in a short article in *Nature* in April 1953, consists of two helices would around each other like a kind of spiral staircase with steps composed of paired bases, and with the sugar-phosphate backbone running along the outside (see Figure 7.8). The two strands of the Watson-Crick model are *complementary* to each other, so that wherever an adenine appears on one strand, a thymine occurs directly opposite from (and is hydrogen bonded to) it in the other strand.

This model accounted beautifully for the major genetic, biochemical, and structural characteristics of the hereditary material. On the biochemical and

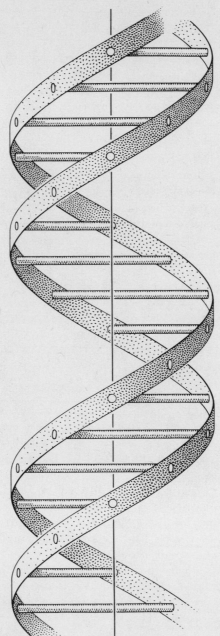

Figure 7-8   DNA Helix (from Nature 171 (1953): 737-8.)

structural level, it explained the x-ray data characteristic of helices, the constant diameter of the fiber, the stacking of the bases at regular intervals, and the 1:1 base ratios. On the biological level, it explained both auto-catalysis and heterocatalysis and suggested a mechanism for how DNA stores genetic information. Autocatalysis, or the replication of the DNA molecule, could occur by each strand acting as a template for constructing its partner. Thus, when the strands separated, each formed its complement, the result being two complete molecules where only one existed before.

Although no proof existed for the mechanism of heterocatalysis in 1953, the new model made it possible to envision how DNA could function to guide the synthesis of other molecules. Each DNA strand could act as a template to form an intermediate message-carrying molecule, presumably one of the other types of nucleic acids, RNA (ribonucleic acid). RNA produced from the DNA template could then carry the genetic information from the nucleus to the cell's metabolic machinery in the cytoplasm, where it could guide the formation of specific proteins. Probably more important, however, was Watson and Crick's idea that the sequence of bases (nuc-leotides) along any one strand of the DNA molecule contained specific genetic information, which became translated into a determined sequence of amino acids on a protein. Thus the old idea of the biochemical geneticists, that genes produce proteins, or polypeptides, could now be understood in terms of the sequence of bases along DNA determining the sequence of amino acids along proteins. Although the specific mechanism of coding was not yet known, a direction for studying the problem was at last available.

From 1953 to 1963 the bulk of the answers to these questions was worked out in rapid and almost explosive succession. The existence of several species of RNA, postulated in the late 1950s, was confirmed in the early 1960s. It thus became possible to construct a complete biochemical scheme for gene protein. The same scheme also provided a consistent explanation for DNA replication. The following diagram, schematizing these two sets of relationships, became known as the "central dogma" of molecular biology.

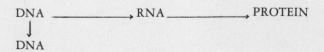

DNA ⟶ RNA ⟶ PROTEIN
↓
DNA

The vertical direction indicates autocatalysis, called by its newer names replication or transmission. The horizontal sequence, indicating heterocatalysis, could actually be divided into two separate reactions, as indicated above. The first (DNA→RNA) is called transcription, and the second (RNA→PROTEIN) is called translation. As the biochemical pathways by which these processes operate became more clear, it was possible to synthesize complete proteins in test tubes with no cells present (using only precursors and DNA or RNA templates). However exciting the *in vitro* synthesis of proteins from nucleic acids may have been, probably the most revolutionary development during the 1960s was the working out of the genetic code in 1963. Independently, in the laboratories of Severo Ochoa (1909-   ) and Marshall Nirenberg (1927-   ) in the United States, sophisticated techniques showed that specific sequences of three bases on DNA (or on the corresponding "messenger" RNA) coded for each of the 20 amino acids. The code was a triplet: for every three bases on DNA, there was a corresponding amino acid in the polypeptide chain. The secret of life, which Watson and Crick had hoped to find, appeared to reside in the informational code of DNA molecules. In plain molecular and biochemical terms, the answer to the question, "How do genes replicate and carry information?", had been found. It had come about largely by the use of information taken from biochemistry (pathways of protein synthesis, Chargaff's work on DNA), studies of molecular structure (x-ray crystallography), and studies on the genetic nature of information-carrying processes in heredity (Delbrück, Hershey, and Chase).

Once the genetic code had been cracked, the dogmatic phase of molecular genetics gave way to what has been called the academic (beginning about 1963 and still continuing today). This final phase was characterized by the working out of myriads of details of the replicative, transcriptive, or translational processes. It was the full-fledged period of what Thomas Kuhn calls "normal science," where the many ramifications of new theories are made complete and fit into a logical system. In particular, the academic phase saw the working out of the molecular structure of transfer RNA, delineation of fine structure of genes and development of the idea of colinearity between DNA and protein, elucidation of the specific structure of a number of proteins, and demonstration of the relationship between gene mutation and structural irregularity in a corresponding polypeptide chain.

The academic phase was also characterized by the failure of Delbrück's and Schrödinger's "other laws of physics" to materialize. The working of the entire central dogma could be accounted for in known physical and chemical terms, down to the various attractive forces holding the double strands of the DNA molecule together. The original hope of the founding fathers of the informationist school, that biology would lead to a new concept of physics, was not realized. In fact, just the opposite occured. Biologists now became more convinced than in the days of Jacques Loeb that the most complex biological phenomena could be understood in terms of the current concepts of physics and chemistry. With the dogmatic and academic phases there was introduced a new wave of mechanistic thinking, and many of the old battle lines were seen to start forming once again. Reductionism, albeit of a somewhat more sophisticated and substantial form thatLoeb had introduced, returned once more to certain areas of biology. This reductionism was more short-lived than its counterpart in the 1920s, since the new spirit of holistic mechanism, both from physics and from areas of biology other than biochemistry, began to enter into the views of molecular geneticists. The existence of complex control mechanisms and the recognition that *in vivo* gene transcription and translation is inordinately more involved than the systems studied *in vitro,* suggested that the biological properties of genetic processes could not simply be reduced to molecular structure or one-to-one chemical reactions.

The central dogma even extended as far as evolutionary theory. First, it provided a molecular explanation of gene mutation. Mutations were nothing more than the substitution of one or more nucleotides (or bases) in a DNA strand. Like a typographical error, a change in a nucleotide could cause the substitution of one amino acid for another or could produce a "nonsense" region of the messenger that caused an amino acid to be omitted from the polypeptide. A specific example of this was shown conclusively in the early 1960s by Vernon Ingram at MIT; he demonstrated that the difference in one amino acid between sickle-cell and normal hemoglobin was the result of the substitution of a single base in the corresponding section of DNA. Thus one typographical error in replicating the code had enormous consequences of survival for numerous individuals in succeeding generations. This and other examples suggested that even the mechanism of evolution could be taken down to the molecular level and understood in the same terms that accounted for the processes of genetic transmission, transcription, translation, and embryonic differentiation. The unity of biology, sought after for so long, seemed to be finally at hand.

Conclusion

Molecular biology and its specific subdiscipline, molecular genetics, concern both the structure and function of biologically important molecules. Molecular genetics grew out of the attempt to determine the nature of the

gene and how it works. As such its area of concern involved those traditionally limited to genetics or biochemistry. Yet molecular genetics differs from
both of these more limited fields. It differs from genetics which, until the
late 1930s, was concerned primarily with the problem of transmission, by its
concern for the mechanism of gene function. It differs from biochemistry
which, until the 1950s, was concerned primarily with the function of
molecules, by its concern for three-dimensional molecular architecture. The
initial strides in what we now call molecular biology were made in molecular
genetics, whose conclusions and methods were later extended to other areas
such as development, evolution, immunology (study of the immune response), or cell physiology (study of cell membrane structure and function).
By the late 1960s, the molecular approach had invaded many areas of
biology; a truly molecular biology had been born.

Molecular genetics grew out of the attempt to apply functional and
informational questions to the highly elaborate but formalistic structure of
the classical Mendelian chromosome theory. From the disenchanted quantum physicists such as Bohr, Schrödinger, and Delbrück came the romantic
impetus to find the elusive "secret of life." That secret, they were convinced,
came not in simply reducing life to known physical and chemical laws but
in finding new laws of the physical universe from the study of life. The secret
they were hoping to find lay, they thought, in the nature of the gene. In
particular, this meant the means by which information was stored and
transmitted and by which replication of information was accomplished.
Interestingly, the romantics of early molecular genetics belied their quantum
theory background by their antimaterialism and by their disdain for
biochemistry. They admired the formal superstructure of classical genetics,
untainted as it was by any particular chemical considerations. Those who
migrated from physics into biology carried a similar approach. The early
informationists were uninterested in the details of molecular structure. They
wished to pursue classical genetics to what seemed like its own next logical
step: What is the informational content of the "gene?" As phage workers,
their major concern was the transmission of information during phage
infection, not the biochemical pathways of gene action.

The dogmatic phase of informationist theory involved the explicit infu-
sion of biochemical and structural considerations into the more formalistic
work of the early phage group. From biochemical genetics came the explicit
connection between genes and proteins. The direct product of gene action
was a polypeptide chain, which formed all or at least part of a whole protein
molecule. From the structurists' work came, first, the methodology of
deducing the three-dimensional structure of large molecules by analysis of

x-ray diffraction patterns. Second came the notion of three-dimensional specificity of proteins as a function of amino acid sequence. To a younger generation of informationists, who were both less enamored of classical genetics and less skeptical of biochemsitry, these findings suggested a possible colinearity between the sequence of information in the gene and the sequence of amino acids in its product, the protein. To discover this colinearity, it was first necessary to determine what the explicit genetic material was and then to work out its three-dimensional structure. Foremost in the minds of this second phase of the informationist school were not the chemical or physical considerations of molecular structure but the biological. They were guided by a desire to find a molecular structure that would account for the ability of genetic material to replicate itself and to code for amino acid sequences in proteins. That DNA and not protein was the genetic material had not be any means been clear-cut before the early 1950s. And yet the "feeling" among many informationists, especially those who had come under the influence of Delbrück and Luria, was that DNA was the likely candidate. The work of Hershey and Chase in 1952, following that of Avery and his co-workers in 1944, confirmed that feeling. But by this time Watson, Circk, and others (including Pauling) were honing in on nucleic acids.

The work of Watson and Crick brought together the informationist, structurist, and biochemical appraoches to genetic (and, by extension, all biological) problems. They realized that it was essential to know the exact structure of the genetic material, down to every bond angle and all spatial distances between different atoms and groups of atoms. And for this purpose it was necessary to use the approaches developed in the structurist school. Only in this way, they reasoned, could the exact mechanism of gene replication and control of protein synthesis be understood. But gene replication and the guidance of protein synthesis were distinctly biochemical problems—they involved the use of pools of precursors in the cell, and were intimately connected to various metabolic pathways. Moreover, knowledge of the biochemistry of DNA, for example, the specific forms of the bases in the *in vivo* molecule, was crucial to developing the ultimate molecular structure.

Although they saw the importance of using the structural and biochemical information about DNA, Watson and Crick were primarily informationists at heart. Their interest was guided from the start by Schrödinger's intriguing questions about genes and the secret of life. The most important question became, "How does the structure of the genes become transformed into the structure of proteins?" This was the key to understanding life, because this was how life differed from most nonliving phenomena with which scientists were familiar. Structure was important, but it was subservient in Watson's and Crick's minds to function. Every structural feature had to be consistent with the biological demands placed on the DNA molecule. It was this insistence on relating structural and biochemical knowledge to the requirements of biological function that allowed Watson and Crick to put together the formerly separate structurist and biochemical approaches.

It would appear from all that has gone before that molecular genetics represented a sharp break from classical genetics and that the molecular approach to biology as a whole represented a break from the trend toward antireductionism embodied in the work of Henderson, Cannon, and others. Neither is quite the case. In fact, molecular biology represents a strong element of continuity with its past.

Classical genetics of the Morgan school had raised questions about the nature of the gene and gene action that could not be answered in the three decades following the discovery of *Drosophila* in 1910. But the questions were kept alive while work on other issues was undertaken in the laboratory. What Morgan and his co-workers had done was to erect an elaborate, logically consistent, but formalistic concept of genetics that was wholly detached from any biochemical foundation. It was materialistic in drawing the relationship between invisible genes and visible chromosomes, but it lacked completely any concept of chemical or physiological function. Yet the existence of functional questions raised by Morgan and others did not fall on completely deaf ears. It is not coincidental that two leaders in the later biochemical and informational approaches to genetics, George Beadle and Max Delbrück, respectively, worked in Morgan's laboratory at Cal Tech for several years. They carried away with them, first, the realization that functional questions were extemely important. That was, after all, the expressed aim of Morgan's department at Cal Tech: to give emphasis to functional (and that meant physical and chemical) approaches to biology. Second, they learned the enormous intricacy of gene organization—the idea that genes must be extremely large and complex substances. Third, they saw that *Drosophila* was probably not going to be the same favorable organism for functional studies that it had been for studying chromosome structure and the mechanism of transmission. Each set out to find a system that would give better answers to the new questions. Thus, while molecular genetics represented a departure from classical genetics, it nevertheless grew out of the earlier genetic work and, in some ways, carried that work to its logical extension—the molecular level.

Nor did molecular biology, especially in its earliest stages, represent a return to the old-fashioned mechanistic materialism of Jacques Loeb. The émigré physicists who launched the informationist approach to biology—particularly Bohr, Delbrück, and Schrödinger—were themselves anti-mechanists and antireductionists in the older sense of the words. They all admitted that living organisms operated by the same laws as those governing nonliving matter; there was no vital force in the living world that was absent from its nonliving counterparts. But they did feel that the older forms of reductionism obscured basic features of the complexity of life. To try and reduce these phenomena simply to the level of known chemical reactions was to lose sight of the fact that their complex organization was, in fact, a part of their biological characteristics. Delbrück, for example, made his point clear by singling out explicitly the work of Loeb as exemplifying "a naive conception of biology as devoid of its own laws and indistinguishable from the physical sciences." Loeb, he felt, simply wanted to make all biological phenomena a branch of biochemistry. The link that Delbrück and others seemed to discern between biochemistry and naive reductionism may account for their hostility toward this subject—a hostility they conveyed to others engaged in the pioneering phases of the phage work. Neither Delbrück nor others of the early informationist school claimed that no one should study biochemistry or that biochemistry should not be applied to genetic problems. What they did oppose was the idea that biochemistry was the starting and finishing point of every biological investigation. Other approaches were important and must be pursued if a truly meaningful biology was to emerge.

As we saw in Chapter IV, this view was similar to one adopted by Henderson and others and was related to the notions of complementarity and phenomenalism introduced first into physics by the quantum theory. Thus it is not surprising that the early informationist school of biology, composed as it was of former quantum theorists, should have the same philosophical biases. The development of the informationist school thus grew out of the same view of the nature of reality and philosophy of science as Henderson's work on self-regulating systems. Yet the later development of molecular biology did show a tendency toward returning to old-fashioned reductionism. Some phage and *E. coli* workers of the early 1960s were known to make statements about cells as mere collections of molecules, or what "what's true of *E. coli* [a bacterium] is true of the elephant," in a vein similar to Loeb's more outrageous pronouncements. However, many molecular biologists came to realize that knowledge of the complete molecular structure of DNA did not necessarily offer knowledge about its function, its self-regulatory processes (how does it know when to turn on or off?), or its evolution. The lessons of the early informationists have carried down through the development of modern molecular biology. A tendency to revert to an older mechanistic view, encouraged by the major successes of the molecular approach in the late 1950s and early 1960s was thus restrained by the persistent philosophy of those who had founded the field two decades earlier.

The growth of molecular biology has shown the union of evolutionary, hereditary, embryological, biochemical, and even anatomical (in a new guise, as molecular anatomy) concepts in a manner that would have delighted the hearts of Ernst Haeckel and August Weismann. It was a true union, because it took a whole host of phenomena down to the lowest common denominator and, to biologists of the twentieth century, this meant molecules. The central ideas of modern molecular biology have shown the same kind of unifying effect on the study of life as Darwin's did a century earlier. Yet the criteria of theory building and of seeking certain types of evidence were wholly different in the two periods. Molecular biology saw the culmination of a trend, first begun in the 1880s by Roux's *Entwicklungsmechanik*, of treating the organism in an experimental way, using the tools of physics and chemistry. Whereas Darwin's theory did not suggest any immediate ways in which it could be experimentally tested (in fact, it was thought that evolution was too slow to ever be observed in the laboratory), the basic tenets of the central dogma could be (and were!) almost immediately put to the test. Predictions from Darwin's theory could not be verified, whereas those from the central dogma could be easily verified (or contradicted, as the case may be). Thus, while both theories provided a focus for unifying a number of disparate areas of biology, they achieved this end by wholly different methodologies. It is this change in methodology which has characterized the growth of life sciences from the late nineteenth century through the present time.

Bibliography

BIOLOGY AND ITS NINETEENTH-CENTURY BACKGROUND

There are no books dealing in a direct and general way with the history of any of the sciences, but especially biology, in the twentieth century. Erik Nordenskiold's *The History of Biology* (New York, 1928), despite its age, is still one of the few sources dealing with biological history in the early years of the century. Nordenskiold gives sketchy treatment to what he does discuss, however, and is of limited value. A variety of topics is discussed in R. Harré's *Scientific Thought 1900-1960: A Selective Survey* (Oxford, Clarendon, 1969), dealing mostly with chromology and the facts of who discovered what and when. The authors of the individual essays are specialists in one or another scientific discipline who chronicle the development of discoveries and ideas in their fields. There are essays on "Biochemistry" by S. G. Waley, "Molecular Biology" by G. H. Haggis, "Ecological Genetics" by E. B. Ford, "Hormones and Transmitters" by H. Blaschko, "Cell Biophysics" by D. Nobel, "The Viruses" by N. W. Pirie, and "Ethology" by N. Tinbergen. Some useful insight into areas such as genetics, evolution, behavior, and natural history can be gleaned from Mary Alice and Howard Ensign Evans' *William Morton Wheeler, Biologist* (Cambridge, Mass., 1970). More than a biography, this well-written book is in many ways a portrait of the shift from natural history dominated to laboratory-dominated biology.

For the nineteenth-century background to twentieth-century biology, William Coleman's *Biology in the Nineteenth Century* (New York, 1971) is the best starting place. John Theodor Merz' *A History of European Scientific Thought in the Nineteenth Century* (New York, reprint edition, 2 vols., 1964), originally published between 1904 and 1912, is a comprehensive and detailed account of the sciences in general and their relation to philosophy. Most of the valuable information on biology is found in Volume II (pp. 200-464). A less thorough but more recent work is Ben Dawes' *A Hundred*

Years of Biology (London, Duckworth, 1952), which covers some of the early years of the twentieth century. For enunciation of the general idea that biology first became solidified as a comprehensive science in its own right during the nineteenth century, see J. W. Wilson, "Biology attains maturity in the nineteenth century," in Marchall Clagett, ed., *Critcial Problems in the History of Science* (Madison, Wisconsin 1959): pp. 401-418; Wilson emphasizes particularly the role of the cell and protoplasm theories as unifying concepts in mid- and late nineteenth-century biological thought. A recent book by Elizabeth Gasking, *The Rise of Experimental Biology* (New York, Random House, 1970) discusses in a very readable fashion the growth of experimentalism from the seventeenth through the end of the nineteenth centuries.

For an understanding of the morphological movement, Patrick Geddes' article "Morphology" in the ninth edition of the Encyclopedia Brittanica (1883), Vol. 16, pp. 837-846, remains one of the most complete and concise accounts of the history and meaning of this important area of nineteenth-century biology. A more expanded treatment can be found in E. S. Russell's *Form and Function*, first published in 1916 and now available in a reprinted edition, with an introduction by William Coleman (New York, 1967). The work of another more recent morphologist, D'Arcy Wentworth Thompson, has been carefully analyzed by Stephen Gould: "D'Arcy W. Thompson and the science of form," *New Literary History, 2,* (1971): pp. 229-258. In regard to the classical morphologists of the late nineteenth and early twentieth centuries, the life and work of August Weismann has been treated by Ernst Gaupp, *August Weisman, sein Leben und sein Werk* (Jena, 1917); this is the most complete and useful biography. Despite its title, Weldemar Schleip's August Weismann's Bedeutung fur die Entwicklung der Zoologie und allgemeinen Biologie," *Naturwissenschaften, 22* (1934): 34-41, does not devote much space to Weismann's influence on later developments in biology—particularly in the early twentieth century. A more recent account, containing much valuable information, both personal and scientific, is Helmut Risler, "August Weismann (1834-1914)," *Berichte Natur-forschung Gesellschaft Freiburg in Bresigau, 58* (1968): 77-93. The most authoritative biography of Haeckel is Heinrich Schmidt's *Ernst Haeckel: Leben und Werke* (Berlin, 1926); the philosophical side of Haeckel's work is emphasized in a more recent study by David DeGrood, *Haeckel's Theory of the Unity of Nature: A Monograph in the History of Philosophy* (Boston, 1965). The history of the biogenetic law (recapitulation theory) is the subject of A. W. Meyer's, "Some historical aspects of the recapitulation idea," *Quarterly Review of Biology, 10* (1935): 379-396; of J. H. F. Kohlbrugge's, "Das biogenetische Grundsetz. Eine historische Studie," *Zoologische Anzeiger, 38* (1911): 447-453; and more recently of Jane Oppenheimer's, "Embryology and evolution: nineteenth century hopes and twentieth century realities," *Quarterly Review of Biology, 34* (1959): 271-277, reprinted in Oppenheimer, *Essays in the History of Embryology and Biology* (Cambridge, Mass., 1967): 206-220. The work and influence of one of the chief American morphologists, W. K. Brooks, is treated by Dennis McCullough, "W. K. Brook's role in the history of American biology," *Journal of the History of Biology, 2* (1969): 411-438.

The fate of the Darwinian theory in the post-1860 period has not been treated extensively, especially when compared with the quantity of written material dealing with the pre-1860 period. However, the following give some indication of the course that late nineteenth-century evolutionary theory followed: two contemporary accounts of the growing *scientific* opposition to Darwin between 1880 and 1900 are Yves Delage and Marie Goldsmith, *The Theories of Evolution* (New York, 1912), and V. L. Kellogg, *Darwinism Today* (New York, 1907); the latter is an exceptionally rich reference source and summary of the primary literature. Some aspects of this period are treated in Loren Eisley's fine book, *Darwin's Century* (New York, 1961), especially Chapter 9, "Darwin and the Physicists." Two essays by Garland Allen detail the reception of and objections to Darwinian theory in the early twentieth century: "Thomas Hunt Morgan and the problem of natural selection," *Journal of the History of Biology, 1,* (1968): 113-139, and "Hugo de Vries and the reception of the 'Mutation Theory,' "*Journal of the History of Biology, 2* (1969): 55-87. An insight into some of the complexities of early twentieth-century evolutionary theory is found in Stephen Gould's "Dollo on Dollo's Law: Irreversibility and the status of evolutionary laws," *Journal of the History of Biology, 3* (1970): 189-212. Neo-Lamarckism in the early twentieth century has been investigated by Edward J. Pfeifer in "The genesis of American neo-Lamarckism," *Isis, 56* (1965): 156-167; some indication of the extent to which biologists sought evidence to justify neo-Lamarckism can be found in Arthur Koestler's *The Case of the Midwife Toad* (London, 1971), a study of the life and work of Paul Kammerer. Koestler's book must be approached more as an apology for Kammerer and, to some extent, his ideas of the inheritance of acquired characters, than as an accurate historical assessment of the neo-Lamarckian movement. The religious response to Darwinian theory in twentieth-century culture has been simply and accurately captured in John Scopes' personal memoir, *Center of the Storm* (New York, 1967). Ernst Benz's *Evolution and Christian Hope,* trans. by H. G. Frank (Garden City, N. Y., 1966) and John Dillenberger's *Protestant*

Thought and Natural Science (Garden City, N. Y., 1960) deal with the same problem in a more general way. The influence of evolutionary views on the social sciences is discussed in Richard Hofstadter's *Social Darwinism in American Thought* (Boston, 1955), the best available treatment of the Darwinian ethic as it came to be a rationalization for capitalist exploitation. George W. Stocking has discussed the influence of specifically neo-Lamarckian views on social theory in "Lamarckianism in social sience: 1890-1915," *Journal of the History of Ideas, 23* (1962): 239-256. Scientific attitudes on the evolution of race have been thoroughly detailed in John S. Haller, *Outcasts from Evolution: Scientific Attitudes of Racial Inferiority 1859-1900* (Chicago, 1971).

The mechanism-vitalism debates in the nineteenth century are treated in D. C. Phillips, "Organicism in the late nineteenth and early twentieth centuries," *Journal of the History of Ideas, 31* (1970): 413-432; Gerald Geison, "The protoplasmic theory of life and the vitalist-mechanist debate," *Isis, 60* (1969): 273-292; and in Reinhard Mocek's penetrating "Mechanismus and Vitalismus," *Mikrokosmos-Makrokosmus, 2* (1967): 324-372. More specific references on the mechanistic view in twentieth-century biology will be found in the bibliography for Chapter IV.

Several works dealing with philosophy of biology, in particular the nature of scientific explanations, are especially helpful. Gwynn Nettler's *Explanations* (New York, 1970) is a very general introduction to the nature of explanation, drawing examples from all disciplines and walks of life. It is a readable and down-to-earth presentation. For an understanding of specifically scientific modes of explanation, see Arthur Pap, "What is a law of nature?," Thomas Kuhn, "Paradigms and some misinterpretations of science," and Mary Hesse, "The role of models in scientific theory," all in Dudley Shapere (ed.), *Philosophical Problems of Natural Science* (New York, 1965). Michael Scriven's "Explanation in the biological sciences," *Journal of the History of Biology, 2* (1969): 187-198, along with most of the articles in that issue (Volume 2, Number 1), is concerned primarily with biological examples. Among the other articles that may be of interest are Dudley Shapere's "Biology and the unity of science" (pp. 3-18), Kenneth Schaffner's "Theories and explanations in biology" (pp. 19-34), and Richard Lewontin's "The bases of conflict in biological explanation" (pp. 34-46). Ernst Mayr's "Cause and Effect in biology," *Science, 134* (1961): 1501-1506 is a thought-provoking essay differentiating the types of explanations often used by biologists for different types of phenomena. More recently David Hull's *The Philosophy of Biological Science* (Englewood Cliffs, N.J., Prentice-Hall, 1974) summarizes in a cogent form some of the basic problems in modern philosophy of science with respect to biology. Hull's work assumes little formal background to either biology or philosophy; the short space of 140 pages makes the treatment abrupt in places, highly compact throughout, but nonetheless extremely valuable.

With regard to the philosophical concepts of mechanism, idealism, and reductionism in natural science, one of the most useful, and brief, sources is Maurice Cornforth's *Materialism and the Dialectical Method* (New York, 1968). This book is a simple explication of materialism and idealism, with examples of how each view would approach certain concrete problems. The historical side of mechanism in biology in the nineteenth century is treated by Everett Mendelsohn in "Physical models and physiological concepts: Explanation in nineteenth century biology," *British Journal for the History of Science, 2* (1965): 201-219. The nature of mechanistic versus holistic (not vitalistic) explanation in biology from the focus of two papers in Ronald Munson (ed.), Man and Nature (New York, 1971): Ernest Nagel, "Mechanistic explanation and organismic biology (pp. 19-32); and Morton Beckner, "Organismic biology," (pp. 54-62). Kenneth Schaffner's articles, "Approaches to reduction," *Philosophy of Science, 34* (1967): 137-147 deals with the philosophical issues involved in "reducing" complex phenomena to their component parts, and the types of explanation that emerge from this process. A treatment of the mechanism-vitalism controversy in the twentieth century is found in Hilde Hein's "The endurance of the mechanism-vitalism controversy," *Journal of the History of Biology, 5* (1972): 159-188.

The philosophy of evolutionary theory has received more attention among contemporary philosophers of science that any other single problem in the history of biology. An entire section (III) of Munson's anthology has been devoted to philosophical issues in evolutionary theory with articles by Marjorie Grene, Leigh van Valen, Theodosius Dobzhansky, George Gaylord Simpson, Ernst Mayr, and John R. Gregg among the most interesting and valuable.

REVOLT FROM MORPHOLOGY I: THE ORGINS OF EXPERIMENTAL EMBRYOLOGY

Among the best accounts of some specific problems in late nineteenth- and early twentieth-century embryology are those by Jane Oppenheimer, herself a practicing embryologist with a persistent interest in history. Several of her most useful writings, now collected in her book, *Essays in the History of Embryology and Biology* (Cambridge, Mass., 1967), include: "Embryological concepts in the twentieth cnetury" (pp. 1-61); "Questions posed by classical descriptive and experimental embryology" (pp. 62-91); "Ross Harrison's contributions to experimental embryology" (pp. 92-116); "Analysis of development: Problems, concepts and their history" (pp. 117-172), and "Analysis of development: Methods and techniques" (pp. 172-205); the latter two essays are taken from B. H. Willier, Paul Weiss, and Viktor Hamburger (eds.), *Analysis of Development* (Philadelphia, 1955). A general survey of nineteenth-century embryological developments is found in William Coleman's *Biology in the Nineteenth Century* (op. cit.), Chapter III.

The papers of Wilhelm Roux and Hans Driesch, outlining their classic experiments, are both translated and reprinted in the excellent little volume edited by B. H. Willier and Jane Oppenheimer, *Foundations of Experimental Embryology* (Englewood Cliffs, N. J., 1964); Roux's "Contributions to the developmental mechanics of the embryo. On the artificial production of half-embryos by destruction of one of the first two blastomeres, and the later development (post generation) of the missing half of the body" (pp. 2-37), and Driesch's "The potency of the first two cleavage cells in echinoderm development. Experimental production of partial and double formations" (pp. 38-50) are reprinted in full. Parts of Driesch's paper are also reprinted in M. L. Gabriel and Seymour Fogel, *Great Experiments in Biology* (Englewood Cliffs, N. J., 1955), pp. 210-214. Excerpted versions of both the Roux and Driesch papers appear in Thomas Hall's *A Source Book in Animal Biology* (New York, 1961), Roux, pp. 412-418; Driesch, pp. 418-426. The introduction to volume one of Roux' *Archiv fur Entwicklungsmechanik der Organismen* (1894-1895), in which the philosophy of *Entwicklungsmechanik* is outlined, has been translated by William Morton Wheeler as "The problems, methods and scopes of developmental mechanics," in *Biological Lectures Delivered at the Marine Biological Laboratories of Woods Hole in the Summer Session of 1895* (Boston, 1895), pp. 149-190. No modern studies of the work of Roux or Driesch have yet been published, although Frederick Churchill's Ph.D. dissertation, *Wilhelm Roux and a Program for Embryology* (Cambridge, Mass. 1968) is an important step in rectifying that situation. Churchill's recent article, "Chabry, Roux, and the experimental method in nineteenth century embryology," in R. N. Giere and R. S. Westfall, *Foundations of Scientific Method: The Nineteenth Century* (Bloomington, Indiana, 1973), presents a clear picture of the influences operating on Roux. A series of articles appeared in *Die Naturwissenschaften, 238* (1920): 429-459 as part of a testimonial on the occasion of Roux's seventieth birthday. Two of particular interest are Hans Spemann's "Wilhelm Roux als Experimentator," pp. 443-445 and Hans Driesch's, "Wilhelm Roux als Theoretiker," pp. 446-450. Biographical details are given in Dietrich Barfurth's introductory essay, "Wilhelm Roux zum siebzigsten Geburtstage," pp. 431-434. Barfurth wrote a somewhat larger biographical sketch after Roux's death, "Wilhelm Roux, ein Nachruf," *Archive für Mikroskopisch Anatomie und Entwicklungsmechanic, 104* (1925): i-xxii; a more recent study of Roux as both an anatomist and philosopher has been made by Reinhard Mocek, "Wilhelm Roux (1850-1924)—Anatom und Philosoph," *Wissenschaftliche Zoologie Universität Halle, 14* (1967): 207-222. Mocek has also written two recent papers on Hans Driesch, discussing primarily his philosophical and political, as opposed to scientific, ideas: "Engagement fur Frieden und Humanismus: Gedachtnis-kolloquium anlasslich des 100. Geburtstages von Hans Driesch," *Deutsche Zeitschrift für Philosophie, 16* (1968): 353-360; and "Hans Driesch als politischer Denker," *Schriftenreihe fur Geschichte der Naturwissenchaften, Tecknik, und Medizin, 4* (1967): 138-150. A fine study of the philosophical positions implicit in Driesch's experiments and the traditions they inaugurated is found in Frederick Churchill's "From machine-theory to entelechy: two studies in developmental teleology," *Journal of the History of Biology, 2* (1969): 165-185.

Some impressions of the excitement rampant during the early years of developmental mechanics can be found in Thomas Hunt Morgan's article "Developmental mechanics," *Science, 7* (1898): 156-158. Particularly interesting is the role of several marine laboratories, notably at Naples and at Woods Hole, Mass. in fostering the new experimental embryology. Morgan wrote a glowing account of his 10 months in Naples in 1894-1895: "Impressions of the Naples zoological station," *Science, 3* (1896): 16-18. Some interesting material regarding the MBL if sound in F. R. Lillie's history, *The Woods Hole Marine Biological Laboratory* (Chicago, 1944), especially Chapter 3, "The founding and early history of the marine biological laboratory," and Chapter 6, "Research at the Marine Biological Laboratory: the first twenty years."

The interrelations between early experimental embryology and cell biology can be gleaned from varied sources. Arthus Hughes' *A History of Cytology* contains much valuable information but stops early into the twentieth century, thus cutting short the story of the convergence that occurred after 1910. Still one of the best sources for the early history of cell and developmental studies is found in the writings of the American cytologist, Edmund Beecher Wilson—in particular his classic volume, *The Cell in Development and Inheritance,* first published in 1896 (subsequent editions in 1900 and 1925); the 1896 edition is now available in a repirnt edition (New York, 1966) with a lengthy introduction by H. J. Muller; see also Wilson's article, "The problem of development," *Science, 21* (1905): 281-294. The work of

Theodor Boveri, which directly inspired Wilson and others to study developmental aspects of the cell, has been discussed in two works: Jane Oppenheimer's "Theodor Boveri: the cell biologist's embryologist," *Quarterly Review of Biology, 38* (1963): 245-249, treats primarily Boveri's scientific work; Fritz Baltzer, in *Theodor Boveri, Life and Work of a Great Biologist, 1862-1915,* translated by Dorothea Rudnick (Berkeley, 1967), provides an integrated study of Boveri's life with considerable detail about his work.

REVOLT FROM MORPHOLOGY II: HEREDITY AND EVOLUTION

Of all the areas of biology in the twentieth century the history of genetics has been the most fortunate in receiving historical treatment. Several general histories of heredity and genetics have been published within recent years. Hans Stubbe's *Kurze Geschichte der Genetik bis zur Wiederentdeckung der Vererbungsregeln Gregor Mendels* (Jena, 1965) is the best source on the period up through the end of the century. L. C. Dunn's *A Short History of Genetic* (New York, 1965) treats the period after 1900 in broad outline, providing many insights into the questions and problems Mendelism raised. A. H. Sturtevant's *A History of Genetics* (New York, 1965) covers similar ground; however, it is less broad in some ways and delves less into historical analysis. L. C. Dunn has also edited an older collection of eassays, *Genetics in the Twentieth Century* (New York 1950), specific articles from which will be cited under appropriate topics below. Curt Stern's shorter essay, "The continuity of genetics," *Daedalus, 99* (1970): 882-908, traces in broad outline the convergence of genetics with cytology and the establishment of the theory of the gene.

The work of Galton, Pearson, and the biometricians has been treated in several portions of books and in a few articles, although nowhere exhaustively. The development of the biometrical movement is the subject of Franz Weiling's "Quellen und Impulse in der Entwicklung der Biometrie," *Sudhoff Archive. Zeitschrift für Wissenschaftsgeschichte, 53* (1969): 306-325. In his recent book, *The Origins of Theoretical Population Genetics* (Chicago, 1971), William Provine devotes two chapters to the biometricians and their subsequent conflict with the Mendelians. The work of Galton, in particular, has been treated in several papers. Galton's law is the subject of a detailed analysis by R. G. Swinburne, "Galton's law—formulation and development," *Annals of Science, 21* (1965): 15-31; the relationship between Galton's view of heredity and evolution is the subject of J. S. Wilkie's, "Galton's contribution to the theory of evolution with special reference to his use of models and metaphors," *Annals of Science, 11* (1955): 194-205, and also of

Ruth Schwartz Cowan's "Sir Francis Galton and the continuity of germ-plasm: a biological idea with political roots," *Actes XII° Congres International d' Histoire des Sciences* (Paris, 1968): 181-186. Robert Olby has discussed Galton's approximation of breeding ratios similar to Mendel's as early as 1875: "Francis Galton's derivation of Mendelian ratios in 1875," *Heredity, 20* (1965): 636.

The work of Mendel has received considerable attention, both in historical analyses of Mendel's own work and the events surrounding its rediscovery in 1900. Mendel's original paper has been translated into English and is available as a separate booklet "Experiments in plant hybridization" (Cambridge, Mass., 1960). Portions of the paper are reprinted in James A. Peters' *Classic Papers in Genetics* (Englewood Cliffs, N.J. ,1964), pp. 1-19, and in J. H. Bennett (ed.), *Experiments in Plant Hybridization* (London, 1965). Mendel's correspondence with botanist Carl von Nägeli contains a clear presentation of most of Mendel's ideas and results; the extant portions of Mendel's correspondence to Nägeli has been translated and republished as "The birth of genetics," *Genetics, 35* (Supplement, No. 5, Part 2, 1950): 1-29. The longest of these letters has been republished in M. L. Gabriel and Seymour Fogel, *Great Experiments in Biology* (Englewood Cliffs, N.J., 1955): 228-239. Robert Olby has discussed Mendel's precursors in the agricultural breeder and naturalist traditions in *Origins of Mendelism* (New York, 1967); Olby also assesses both Mendel's work and the impact of the rediscovery on the first decade of the twentieth century. Much of the current research on Mendel, his life and times, can be gleaned from articles in the periodical, *Folia Mendeliana,* published by the Moravian Museum in Brno, Czechoslovakia; six volumes of this journal have appeared to date. Some of the most interesting papers can be found in a recent volume, containing the proceedings of the Gregor Mendel Colloquium, held in Brno from June 29 to July 3, 1970. Further papers are found in an earlier symposium held in Prague in 1965, M. Sosa (ed.) *Gregor Mendel Memorial Symposium, 1865-1965* (Prague, 1966). The *American Philosophical Society* also produced a similar symposium in 1966; the results were published as a number of the *Proceedings of the American Philosophical Society, 109* (1965). Of general interest among the latter is L. C. Dunn's article, "Mendel, his work, and his place in history," (pp. 189-198). The neglect of Mendel during his own lifetime has been a source of contant historical interest. Elizabeth Gasking's paper, "Why was Mendel's work ignored?" *Journal of the History of Ideas, 20* (1959): 60-84 outlines many of the factors possibly contributing to the general indifference that most plant breeders and evolutionists displayed toward Mendel—if they even knew of his work. Alexander Weinstein traces how Mendel's paper was viewed by those of his [Mendel's] contemporaries, who actually knew of it in "The reception of Mendel's papers by his contemporaries," in *Proceedings of the Tenth International Congress of the History of Science* (Ithaca, N. Y., 1962): 997-1001. Conway Zirkle's older paper, "Some oddities in the delayed discovery of Mendelism," *Journal of Heredity, 55* (1964): 65-72 suggests that a certain viewpoint—including statistical thinking—was required before Mendel's ideas of segregation and random assortment could be adequately understood. The factors that may have led to the simultaneous rediscovery of Mendel by de Vries, Correns, and Tschermak are analyzed carefully by J. S. Wilkie in "Some reasons for the rediscovery and appreciation of Mendel's work in the first years of the present century, *British Journal for the History of Science, 1* (1962): 5-18. The idea that Mendel knew what results to expect and the degree to which he selected appropriate data to support his expectation is the focus of a brilliant, if controversial, paper by the late R. A. Fisher: "Has Mendel's work been rediscovered?" *Annals of Science, 1* (1936): 115-137. Fisher's statistical analysis of Mendel's findings has been reexamined in Gavin De Beer's "Mendel, Darwin and Fisher," *Notes and Records of the Royal Society, 19* (1964): 192-226.

The early period in the establishment of Mendelian ideas in the twentieth century has been the subject of a variety of studies. R. C. Punnett's "Early days of genetics," in *Heredity, 4* (1950): 1-10, is an amusing and illuminating account of Bateson's struggle to establish Mendelian theory in England between 1900 and 1906. Punnett, Bateson's chief helper and colleague at the time, writes with a wit and personal involvement that transcends mere anecdotalism. A similar although less vibrant account of the American scene in 1900 is W. E. Castle's, "The beginnings of Mendelism in America," in L. C. Dunn (ed.) *Genetics in the Twentieth Century* (New York, 1950), pp. 59-76. In the same volume Jay Lush's "Genetics and animal breeding" (pp. 493-510) gives some indication of the degree to which animal breeders were prepared for Mendelian results. A. H. Sturtevant has provided an interesting analysis of the kinds of people who flocked to Mendelism in the early twentieth century, in "The early Mendelians," *Proceedings of the American Philosophical Society, 109* (1965): 199-204. In a very interesting article, "Factors in the development of genetics in the United States: some suggestions," *Journal of the History of Medicine and Allied Sciences, 22* (1967): 27-46, Charles Rosenberg explores the reaction to Mendelism among three groups: practical breeders, doctors, and university biologists. The skepticism greeting Mendel's work from several sectors of the biological community between 1900 and 1910 are discussed in the following: Garland Allen, "Thomas Hunt Morgan and the problem of sex determination," *Proceedings of the*

American Philosophical Society, 110 (11966): 48-57; and Wilma George, "The reaction of A. R. Wallace to the work of Gregor Mendel," *Folia Mendeliana,* 9 (1970): 173-178.

Of chief importance in the twentieth century was the confluence of Mendelian breeding analysis with cytological studies, resulting in establishment of the chromosome theory of inheritance (the theory of the gene). The cytological background of this confluence has been treated by various authors. L. C. Dunn's interesting essay "Ideas about living units, 1864-1909, a chapter in the history of genetics," *Perspectives in Biology and Medicine,* 8 (1965): 335-346 (also appearing as Chapter 3 of his *Short History of Genetics*) traces particulate thinking about heredity in the period during which Mendel's work lay unnoticed. Background to the cytological study of chromosomes, cell division, reduction division, and cell reproduction can be found in three detailed and valuable essays: Dealing with the earlier nineteenth century is John R. Baker's "The Cell Theory: a restatement, history and critique. Part V. The multiplication of nuclei." *Quarterly Journal of Microscopical Science,* 96 (1955): 449-481; Frederick Churchill's "Hertwig, Weismann, and the meaning of reduction division circa 1890," *Isis,* 61 (1970): 429-457, deals primarily with the problem of meiosis (reduction division) during germ cell (egg and sperm) formation in the later nineteenth century when chromosomes were accepted as having some bearing on heredity; William Coleman's "Cell, nucleus, and inheritance: an historical study," *Proceedings of the American Philosophical Society, 109* (1965): 124-158, traces, during the middle nineteenth century, the idea that the cell nucleus, and chromosomes in particular, control heredity. Recognition between 1900 and 1905 of the relationship between Mendelian genes and cytological observations of chromosome behavior has been treated in Victor McKusick's "Walter S. Sutton and the physical basis of Mendelism" *Bulletin of the History of Medicine,* 34 (1960) 487-497. More details of the scientific problems involved, both conceptual and technical are found in H. J. Muller's "Introduction" to the reprint edition of E. B. Wilson's *The Cell in Development and Inheritance* (op.cit.), and in his earlier essays, "Edmund B. Wilson—an appreciation, *American Naturalist,* 77 (1943): 142-172.

The development of the chromosome theory of heredity, and opposition to it, has not been thoroughly enough studied, given its importance in twentieth-century biology. Both Dunn and Sturtevant in their *Histories* treat this issue in some detail but, of necessity, gloss over many important concepts and often neglect the controversial and tentative ideas that emerged along the way. The epoch-making book, *The Mechanism of Mendelian Heredity* by Morgan, Sturtevant, Muller, and Bridges, originally published in 1915, has been reprinted (New York, 1972) and gives a good picture of the kinds of problems being attacked by members of the "Drosophila school" between 1910 and 1915. A brief survey of the work of Morgan and his group is treated in Garland Allen's "Introduction" to this reprint edition. H. J. Muller's article, "The development of the gene theory," in L. C. Dunn (ed.) *Genetics in the Twentieth Century* (op. cit.) outlines the basic development of gene and chromosome theory between 1915 and 1950. Elof Carlson's *The Gene: A Critical History* (Philadelphia, 1966) contains a wealth of information on specific, technical aspects of the development of the chromosome and gene theories. It is perhaps the best introduction to the primary literature of the period between 1900 and 1955. Some important papers in classical genetics are contained in Bruce R. Voeller, *The Chromosome Theory of Inheritance* (New York, 1968) and in James Peters' *Classic Papers in Genetics* (op. cit.). Studies of opposition to the chromosome theory can be found in R. G. Swinburne, "The presence-and-absence theory," *Annals of Science, 18* (1962): 131-145; William Colemean, "Bateson and chromosomes: conservative thought in science," *Centaurus, 15* (1970): 228-314; and Garland Allen, "Richard Goldschmidt's oppositon to the Mendelian-chromosome theory," *Folia Mendliana, 6* (1970): 299-303.

The relationship of classical Mendelian genetics to human genetics has been discussed by Curt Stern in "Mendel and human genetics," *Proceedings of the American Philosophical Society, 109* (1965): 216-226; by L. C. Dunn in "Cross currents in the history of human genetics," *Eugenics Review, 54* (1962): 69-77; and by Charles Rosenberg in "Charles Benedict Davenport and the beginning of human genetics," *Bulletin of the History of Medicine, 35* (1961): 266-276. An extension of concern with human genetics during earlier twentieth century was the eugenics movement, popular between 1890 and 1920. This subject has been treated in the following two articles by Kenneth Ludmerer: "American geneticists and the eugenic movement: 1905-1935," *Journal of the History of Biology, 2* (1969): 337-362; "Genetics, eugenics and the Immigration Restriction Act of 1924", *Bulletin of the History of Medicine, 46* (1972): 59-81; by Mark Haller, "Heredity in progressive thought," *Social Science Review, 37* (1963): 166-176, and more thoroughly in his book, *Eugenics: Hereditarian Attitudes in American Thought* (Rutgers, N. J., 1963); and in two different studies of the eugenic views of H. J. Muller: T. M. Sonneborn, "H. J. Muller, crusader for human betterment," *Science, 162* (1968): 772-776; and Garland Allen, "Science and society in the eugenic thought of H. J. Muller," *BioScience, 20* (1970): 346-353.

Philosophical aspects of the rise of classical Mendelian genetics, especially in the work of the *Drosophila* school, has been treated in two articles by Edward Manier: "Genetics and the philosophy of biology," *Proceedings of the American Catholic Philosophical Association* (Washington, D. C., 1965): 124-133; and "The experimental method in biology: T. H. Morgan and the theory of the gene," *Synthese, 20* (1969): 185-205.

MECHANISTIC MATERIALISM AND ITS METAMORPHOSIS: GENERAL PHYSIOLOGY 1900-1930

Application of the mechanistic philosphy to biology in the early years of the century can be seen best in Jacques Loeb's famous book of 1912, *The Mechanistic Conception of Life,* now available in a reprinted edition (Cambridge, Mass., 1964), with an "Introduction" by Donald Fleming. The introduction does much to set Loeb's book in its historical setting and to trace the development of his philosophy. Arnold E. S. Gussin has treated Loeb's tropism theory in "Jacques Loeb: The man and his tropism theory of animal conduct," *Journal of the History of Medicine and Allied Sciences, 18* (1963): 321-336, and Nathan Reingold has given an indication of Loeb's philosophical interrelationships with his contemporaries in "Jacques Loeb, the scientist, his papers and his era," *Library of Congress Quarterly Journal of Current Acquisitions, 19* (1962): 119-130. The influence of Loeb's views on the development of a mechanistic and experimentally oriented biology, particularly in the United States, is found in Garland Allen's "T. H. Morgan and the emergence of a new American biology," *Quarterly Review of Biology, 44* (1969): 168-188. Two biographical sketches of Loeb are also useful: Simon Flexner's "Jacques Loeb and his period," *Science, 66* (1927): 332-337 is general, while W. J. V. Osterhout's "Jacques Loeb," *National Academy of Science Biographical Memoirs, 13* (1930): 318-401 contains more specific details as well as a complete bibliography.

There is much need of detailed historical studies in the history of neurophysiology. The most general coverage is given in Mary A. B. Brazier's lengthy article, "The historical development of neurophysiology," in J. Field, H. W. Magoon, V. E. Hall (eds.) *Handbook of Physiology* (Baltimore, 1960), Section I, Volume I: pp. 1-57. This article contains many references but, because of the amount of material covered, the treatment of any one topic is necessarily limited. Fascinating reading as well as a detailed analysis of the early years of the neuron theory is provided in Ramón y Cajal's autobiography, *Recollections of My Life,* translated by E. Horne Craigie with the assistance of Juan Cano (Cambridge, Mass., 1966, originally published by the American Philosophical Society in 1937). Some specific aspects of the neuron-nerve net controversy have been discussed by Susan Billings, "Concepts of nerve fiber development 1839-1930," *Journal of the History of Biology*, 4 (1971): 275-306. The work of Sherrington has been analyzed in considerable detail by Judith Swazey in *Reflexes and Motor Integration; Sherrington's Concept of Integrative Action* (Cambridge, Mass., 1969); Dr. Swazey has also discussed the problem of cerebral localization as one aspect of growing concepts of neuronal integration in "Action propre and action commune: the localization of cerebral function," *Journal of the History of Biology*, 3 (1970: 213-234. The most detailed biography of Sherrington is Ragnar Granit's *Charles Scott Sherrington, An Appraisal* (London, 1967). Sherrington's philosophy of physiological activity and of scientific and biological methodology is presented in his Gifford Lectures, *Man, On His Nature* (Cambridge, England, 1951).

Although Walter B. Cannon's neurophysiological researches largely have been ignored to date by historians of science, his popular book, *The Wisdom of the Body*, has long been available in paperback (New York, 1969; originally published 1932). Among Cannon's own more technical writings, the following provide an introduction to his basic ideas and techniques: *Bodily Changes in Pain, Hunger, Fear and Rage* (New York, 1929); "Organization for physiological homeostasis," *Physiological Reviews, 9* (1929): 399-431; "The autonomic nervous system, and interpretation," *The Lancet, 1* (1930): 1109-1115; and his famous series of articles, "Studies in homeostasis in normal, sympathectomized and ergotaminized animals," *American Journal of Physiology, 104* (1933): 172-183; 184-189; 190-194; 195-203. A brief biographical account of Cannon is found in R. M. Yerkes' "Walter Bradford Cannon, 1871-1945," *Psychological Review, 53* (1946): 137-146. L. J. Henderson's views on physiological regulation are contained in his Silliman Lectures, *Blood: A Study in General Physiology* (New Haven, 1928). His other interesting, and somewhat more provocative work, *The Fitness of the Environment* (originally problished in 1913) has been made available in paperback (New York, 1958). George Wald's introduction to this reprint edition gives some resumé of Henderson's life and times. Cannon's biographical sketch "Lawrence Joseph Henderson, 1878-1942," *National Academy of Sciences Biographical Memoirs, 23* (1945): 31-58, gives details of Henderson's career. The most modern and thorough study of Henderson's work in regulatory physiology by an historian is John Parascandola's "Organismic and holistic concepts in the thought of L. J. Henderson," *Journal of the History of Biology, 4* (1971): 63-114.

The generalized problem of physiological regulation has been treated to some extent in the historical literature; the items listed below provide some background for the ideas that developed prior to or simultaneous with the work of Henderson, Cannon, and Sherrington. E. F. Adolph's "Early concepts of physiological regulations," *Physiological Review, 41* (1961): 737-770 is an outstanding general review. Frederic L. Homes' "The *milieu intérieur* and the cell theory," *Bulletin of the History of Medicine, 37* (1963): 315-335, provides an insight into the relationships between Claude Bernard's concept of the regulated internal environment and the cell as the structural and functional unit of life. The presence of regulatory mechanisms has long been recognized in respiratory physiology, as described in John F. Perkins' "Historical development of respiratory physiology," in W. O. Fenn and H. Rahn (eds.), *Handbook of Physiology* (Baltimore, 1964), Section 3, Volume I ("Respiration"): pp. 1-62. Garland Allen has discussed the elucidation of a specific regulatory mechanism in "J. S. Haldane: the development of the idea of control mechanisms in respiration," *Journal of the History of Medicine and Allied Sciences, 22* (1967): 392-412.

THE GRAND SYNTHESIS

Embryology. The early development of the theories of induction and primary organizer is discussed by Jane Oppenheimer in two articles: "Some diverse background for Curt Herbst's ideas about embryonic induction," *Bulletin of the History of Medicine, 44* (1970): 241-250; and Hans Driesch and the theory and practice of embryonic transplantation," Ibid.: 378-382. Spemann wrote an autobiography covering the period up until about 1920, *Forschung und Leben* (Stuttgart, 1943), which contains a wealth of information about the development of his scientific ideas. The best modern interpretation of Spemann's work is Viktor Hamburger's "Hans Spemann and the organizer concept," *Experientia, 25* (1969): 1121-1125—an insightful view by one of Spemann's former students. Of value also is the article by Martin Schnetter, "Die Ara Spemann-Mangold am Zoologischen Institute der Universität Freiburg im Bresigau in den Jahren 1919-1945," *Berichte Naturforschung Gesellschaft Freibrug i. Br. 58* (1968): 95-110. Jane Oppenheimer has analyzed the relationship between the organizer concept and the study of embryology on a cellular level in "Cells and organizers," *American Zoologist, 10* (1970): 75-88; in this article Dr. Oppenheimer focuses particularly on the work of Spemann's students such as Holtfreter. The metamorphosis of chemical embryology in the wake of the organizer theory is the subject of Joseph Needham's reappraisal, "Organizer phenomena after four decades: a retrospect and prospect," in K. R. Dronamraju (ed.), *Haldane and Modern Biology* (Baltimore, 1968): 277-298; the same article appears as a "Forward" ro the third impression of Needham's *Biochemistry and Morphogenesis* (Cambridge, England, 1966). The relationship between genetics and development can be seen from a contemporary's point of view in T. H. Morgan's *Embryology and Genetics* (New York, 1934). A more recent and historical account, if considerably briefer, is Curt Stern's "From crossing-over to developmental genetics," *L. J. Stadler Symposia, 1, 2* (1971): 21-28. A personal account of many developments in embryology since the 1920s is Victor Twitty's *Of Scientists and Salamanders* (San Francisco, 1966). An interesting, if controversial, approach to modern embryology is Ludwig von Bertalanffy's *Modern Theories of Development,* translated and adapted from the original 1933 German edition by J. H. Woodger (New York, 1962). The reader must be cautioned that Bertalanffy is an apologist for neo-vitalism, a viewpoint that permeates his analysis and appraisal of developmental concepts.

Evolution. The evolutionary synthesis has been treated sparsely by historians. William Provine's *The Origins of Theoretical Population Genetics* (op. cit.) provides a general treatment of the developing strands that were woven together as the genetical theory of natural selection in the 1930s. Provine treats Darwin, the biometricians, the later work of Castle and, of course, Wright, Fisher, and Haldane. This book is valuable for the first foray it makes into a complex and virtually unstudied area in the history of modern biology. Of equal value in the same vein are two articles by Mark Adams on the Russian school of population geneticists in the 1920s and 1930s: "The founding of population genetics: contributions of the Chetverikov school, 1924-1934," *Journal of the History of Biology, 1* (1968): 23-39 and "Towards a synthesis: population concepts in Russian evolutionary thought, 1925-1935," Ibid., *3* (1970): 107-129. Both articles present a little-known side of Russian biology in particular, and of the history of population genetics in general in the pre-Lysenko period. The Lysenko controversy has received considerable attention; older books such as Conway Zirkle's *Death of a Science in Russia* (Philadelphia, 1949), or his *Evolution, Marxian Biology and the Social Scene* (Philadelphia, 1959) tend to use the Lysenko issue to illustrate the fate of science under communistic and/or totalitarian regimes. Fortunately, for a better understanding of real history, two recent books place the issue in a more sound perspective: Zhores Medvedev's *The Rise and Fall of T. D. Lysenko,* translated by I. Michael Lerner (New York, 1969) is a highly personal and informative account of Lysenko's influence by a Russian biochemist. More historically analytical, but also less compelling reading, is David Joravsky's *The Lysenko Affair* (Cambridge, Mass., 1970). This thoroughly documented and scholarly treatise relates Lysenko's official acceptance partly to the historical circumstances of Russia's agricultural crises in the 1920s.

THE CHEMICAL FOUNDATION OF LIFE: THE GROWTH OF BIOCHEMISTRY IN THE TWENTIETH CENTURY

Studies in the history of biochemistry are few and far between. Only a handful of books and published papers aid the modern student in understanding the growth of this very important field. The original papers by a number of the workers referred to in this chapter can be found in some of the following collections: Gabriel and Fogel, *Great Experiments in Biology* (op. cit.): pp. 23-53; included here are Buchner's 1897 paper, Keilin's study of cytochromes of 1925, Warburg's 1930 paper on oxidation, and Sumner's paper on crystallization of urease. Herman Kalckar's *Biological Phosphorylations: Development of Concepts* (Englewood Cliffs, N. J., Prentice-Hall, 1969) contains reprints of many important papers in the history of twentieth-century biochemistry. The compilation is too vast to cite specific papers; suffice it to say that this volume provides a useful introduction to the primary literature in a number of areas og biochemistry since 1900, including cellular phosphorylation (i.e., how energy is transferred), energetics of muscle contraction, and regulation of energy metabolism. In the area of phosphorylation and the role of adenosine triphosphate in energy transfers, Fritz Lipmann's *Wanderings of A Biochemist* (New York, Wiley-Interscience, 1971) contains reprints of eight of the author's classic papers on ATP and biosynthesis. The first part of the book contains Lipmann's scientific autobiography, including an account of his days in Meyerhof's laboratory (1927-1930). The essay is intriguing, but disappointing in its lack of personal and historical detail.

The best single secondary source to date on the history of modern biochemistry in Joseph Fruton's *Molecules and Life* (New York, Wiley, 1972). This work is a compendium of information about the history of biochemistry from 1800 to 1960. Fruton's vast knowledge of the literature and his serious attempt to treat the subject historically make this volume a highly significant contribution to the history of biochemistry. *The Development of Modern Chemistry* by Aaron Ihde (New York, Harper and Row, 1964) contains some sections dealing with biochemistry, especially organic and protein chemistry: however, the book is weak on enzymes and metabolic processes. Some valuable essays can be found in a volume edited by Joseph Needham, *The Chemistry of Life, Lectures on the History of Biochemistry* (Cambridge, University Press, 1970). Of particular interest is Malcolm Dixon's "The history of enzymes and of biological oxidations," (pp. 15-37), and Mikulas Teich's "The historical foundations of modern biochemistry,"·(pp. 171-191). Neither essay is, however, very profound or historically penetrating.

Liebig has aroused considerable curiosity among historians of biology; aside from Claude Bernard and Louis Pasteur, he has been the object of more historical study than almost any physiologist in the nineteenth century. His *Animal Chemistry* (1842) has been republished with a lengthly but valuable introduction by F. L. Holmes (New York, Hafner, 1964); Holmes has treated Liebig further in his recent study, *Claude Bernard and Animal Chemistry* (Cambridge, Harvard University Press, 1974); a valuable discourse on Liebig's life and work is found in Holmes' biographical sketch for the *Dictionary of Scientific Biography*, Volume 8: 329-350. Liebig's vitalism is discussed thoroughly in Timothy Lipman's "Vitalism and reductionism in Liebig's physiological thought," *Isis, 58* (1967): 167-185. The general influence of Liebig's school of thought at Giessen is outlined in J. B. Morrell's "The chemist breeders: the research schools of Liebig and Thomas Thomson," *Ambix, 19* (1972): 1-47.

A few interesting monographs provide insight into the history of specific areas of biochemistry. John T. Edsall, himself a protein chemist of long standing, has written several historical essays of value: "Proteins as macro-molecules: an essay on the development of the macromolecules: an essay on the development of the macromolecule concept and some of its vicissitudes," *Archives of Biochemistry and Biophysics*, Supplement 1 (1962): 12-20, outlines the problems encountered by the idea of proteins as definitely structured molecules; "Blood and hemoglobin: the evolution of knowledge of functional adaptation in a biochemical system," *Journal of the History of Biology, 5* (1972): 205-257, discusses the growth of ideas about the interrelation of structure and function in the hemoglobin molecule. The former article is a rather general, summary treatment; the latter a highly detailed analysis of the problems encountered in working out a molecular picture of hemoglobin in the 1920s and 1930s that agreed with its physiological chemistry. Also of great value is the recent article by Robert Kohler, "The background to Otto Warburg's conception of the *Atmungsferment,*" *Journal of the History of Biology, 6* (1973): 171-192. Kohler shows the ways in which Warburg, between 1907 and 1915, developed his membrane theory of respiration. Of more general interest is the same author's "The enzyme theory and the origin of biochemistry," *Isis,64* (1973): 181-196. Like the proceding one, this essay focuses attention primarily on the early years of the present century, roughly 1900 to 1915, principally on the origin of the "new" biochemistry following in the wake of Buchner's and others' work on free-floating enzymes. A final article of some interest in James B. Sumner's personal account of the difficulty he encountered in gaining acceptance for his crystallization process (of urease). His theory ran counter to the still strong (especially in Germany) colloidal school: see "The story of urease," *Journal of Chemical Education* (1937): 255-259.

THE ORIGINS OF MOLECULAR BIOLOGY

In recent years a spate of histories dealing with the revolution in molecular biology and biochemical genetics between 1940 and 1960 have been published.

Among primary sources, Archibald Garrod's classic studies on metabolic deficiencies were published as "Inborn errors of metabolism," *The Lancet, 2* (1908): 1-7; 73-79; 142-148; 214-220. George Beadle and E. L. Tatum's experiments on the genetics of biochemical pathways, "Genetic control of biochemical reactions in Neurospora," *Proceedings of the National Academy of Sciences, 27* (1941): 499-506 is a beautiful and clear example of reasoning and experimental design. The Watson and Crick model of DNA was first published as "Molecular structure of nucleic acids," *Nature, 171* (1953): 737-738. Garrod's papers have not been reprinted; The Beadle and Tatum paper has been reprinted in Peters' *Classic Papers in Genetics* (op. cit.): 166-172, and in Gabriel and Fogel's *Great Experiments in Biology* (op cit.): 273-280; the Watson and Crick paper has been reprinted in Peters, pp. 241-243. Three collections of papers dealing with viral and bacterial genetics have been issued in the past six years: Edward Adelberg (ed.), *Papers on Bacterial Genetics,* 2nd edition (Boston, 1966); Gunther Stent (ed.), *Papers on Bacterial Viruses*, 2nd edition (Boston, 1966); and Geoffrey Zubay, *Papers in Biochemical Genetics* (New York, 1968).

The origins of molecular biology from its physical, chemical, and biological background has formed the subject of several excellent articles. E. A. Carlson makes a case for H. J. Muller's early ideas on the molecular basis of gene action in "An unacknowledged founding of molecular biology: H. J. Muller's contribution to gene theory, 1910-1936," *Journal of the History of Biology, 4* (1971): 149-170. Gunther Stent's "That was the molecular biology that was," *Science, 160* (1968): 390-395 makes the basic distinction between the structurist and informationist schools. Stent himself comes from the informationist viewpoint and exhibits that bias in his presentation. Stent's more recent, *The Coming of the Golden Age; A View of the End of*

Progress (Garden City, N. Y., 1969) is a testament to the author's conviction that molecular biology has reached the end of its development. Eugene L. Hess, in "Origins of molecular biology," *Science, 168* (1970): 664-669, speaks more from the structurist's point of view. Aspects of the structurists' approach are detailed in four articles: Linus Pauling, "Fifty years of progress in structural chemistry and molecular biology," *Daedalus, 99* (1970): 998-1014, John T. Edsall, "Protein as macromolecules: An essay on the development of the macromolecule concept and some of its vicissitudes," *Archives of Biochemistry and Biophysics,* Supplement 1 (1962): 12-20. Both articles discuss the role that studies of protein structure played in developing a conception of three-dimensional shape in macromolecules. In his Harvey Lectures for 1949 W. T. Astbury describes clearly the role that x-ray diffraction played in developing models for protein structures, "Adventures in molecular biology," *Harvey Lectures, 46* (1950): 3-44. These three authors are all working scientists. Robert Olby has tackled the history of molecular biology from the historian's point of view in "The macromolecular concept and the origins of molecular biology," *Journal of Chemical Education, 47* (1970): 168-174.

The history of biochemical genetics per se is less well discussed by either scientists or historians than the history of molecular biology. Two biologists who worked as biochemical geneticists have presented useful essays: Bentley Glass' "A century of biochemical genetics," *Proceedings of the American Philosophical Society, 109* (1965): 227-236 is prosaic, but leads the reader to numerous problems and some of the primary literature. Considerably more readable and lively is George Beadle's "Biochemical genetics: Some recollections," in Gunther Stent (ed.): Phage and the Origins of Molecular Biology (Cold Spring Harbor, N. Y., 1966): 23-32.

The DNA story has had several tellings. Among those workers involved directly, J. D. Watson's *The Double Helix* (New York, 1968) is an outstanding and candid personal memoir. Watson does not claim to write definitive history; nonetheless his book provides a rare insight into not only an important discovery, but the methods and processes of at least some aspects of modern scientific research. Gunther Stent's "DNA," *Daedalus, 99* (1970): 909-937 deals much more with the late 1960s views of the nature of DNA, the work, and the transcriptional and translational process, than on the history itself. Among historians, only Robert Olby has attempted to deal with the development of the DNA idea in an exhaustive and analytical fashion: "Francis Crick, DNA and the central dogma," *Daedalus, 99* (1970): 938-987.

Watson's book evoked a barrage of replies and criticism, some presenting quite different points of view about the origin of the DNA model. Stent's review, "What they are saying about honest Jim," *Quarterly Review of Biology, 43* (1968): 179-184, analyzes all the other reviews, giving interesting explanations about why each reviewer took the position he did. The most spicy and critical review is Chargaff's, "A Quick climb up Mount Olympus," *Science, 159* (1968): 1448-1449. Robert Olby has collected his historical research into a full book, *The Path to the Double Helix* (London, Macmillan, 1974); this is probably the single best account, from the point of view of historical accuracy and objectivity, that is currently available. *Nature* magazine published a series of articles in the April 26, 1974 issue under the general title, "Molecular biology comes of age," *248* (1974): 765-788. It contains articles by Crick, Pauling, Chargaff, Stent, Olby, and Brenner.

Schrodinger's influence on molecular biology in the 1930s is carefully studied by Olby in another essay, "Schrodinger's problem: What is Life?" *Journal of the History of Biology, 4* (1971):119-148; the philosophical issues raised by the new reductionism in the earlier days of molecular biology is analyzed by Kenneth Schaffner in "Antireductionism and molecular biology," *Science, 157* (1967): 644-647. Schaffner argues that not all biological problems can or should be (at this moment in time at least) reduced to physico-chemical explanations; however, he argues this does not mean such problems can never be reduced to such a level. The lesson of modern molecular biology is, Schaffner argues, that reductionism is possible in any area, but does not always provide the most complete or satisfactory understanding of a complex biological phenomenon.

Index

Alcaptonuria, 199
Andrade, E. N. da C. (1887-   ),
    210
Arrhenius, Svante (1859-1927), 76
Astbury, W. T. (1898-   ), 189
Autocatalysis, 209
Avery, O. T. (1877-1955), 208

Back-cross, 52
Bacteriophage, 206
Balfour, F. M. (1851-1882), 21
Barcroft, Joseph (1872-1947), 97
Bateson, William, 1861-1926), 46,
    51
Beadle, George W. (1903-   ), 200,
    226
Beer, Gavin de, 118
Bell, Charles (1774-1842), 82
Bergmann, Carl (1811-1865), 26
Bergmann, Max (1886-1944), 168
Bernal, J. D., 1901-1971), 211
Bernard, Claude (1813-1878), 74,
    82, 84, 97

Biogenic law, 6
Blastomeres, 27
Bohr, Niels (1855-1962), 195
Boveri, Theodor (1862-1915), 57
Bragg, H. W. (1862-1942), 192
Bragg, W. L. (1890-1971), 192
Bridges, C. B. (1889-1938), 61
Brooks, W. K. (1848-1908), 6, 21,
    57
Büchner, Edward (1860-1917), 157
Buffer systems, 96

Cancer, 159
Cannon, Walter B. (1871-1945), 74,
    97, 100, 120
Cartesian nomogram, 99
Castle, W. E. (1867-1962), 65
Chargaff, Erwin (1905-   ), 216
Chase, Martha (1927-   ), 205, 208
Chetverikov, Sergei S. (1880-1959),
    130
Chiasmatypes, 62
Chromatography, 169

253